0339474

Constructive Philosophy

Constructive Philosophy | Paul Lorenzen

TRANSLATED BY
KARL RICHARD PAVLOVIC

The University of Massachusetts Press

Amherst, 1987

Originally published in German.
"Methodisches Denken," "Protologik. Ein Beitrag zum Begründungsproblem der Logik," "Logischen Strukturen in der Sprache," "Logik und Grammatik," "Das Aktual-Unendliche in der Mathematik," "Die klassische Analysis als eine konstruktive Theorie," "Moralische Argumentationen im Grundlagenstreit der Mathematiker," "Wie ist die Objektivität der Physik möglich?" "Das Begründungsproblem der Geometrie als Wissenschaft der räumlichen Ordnung" from Paul Lorenzen, *Methodisches Denken*, © Suhrkamp Verlag Frankfurt am Main, 1968.
"Aufklärung und Vernunft," "Gleichheit und Abstraktion," "Logik und Hermeneutik," "Konstruktivismus und Hermeneutik," "Wie ist Philosophie der Mathematik möglich?" "Zur Rechtfertigung der deduktiven Methode," "Pascals Kritik an der axiomatischen Methode," "Konstruktive Begründung der Mathematik," "Konstruktive und axiomatische Methode," "Zur Definition von 'Wahrscheinlichkeit'" from Paul Lorenzen, *Konstruktive Wissenschaftstheorie*, © Suhrkamp Verlag Frankfurt am Main, 1974.
"Die dialogische Begründung von Logikkalkülen" and "Rationale Grammatik" from *Theorie des wissenschaftlichen Argumentierens*, edited by Carl Friedrich Gethmann, © Suhrkamp Verlag Frankfurt am Main, 1980.
"Politische Anthropologie" and "Geometrie als messtheoretisches Apriori der Physik" from Oswald Schwemmer (Hrsg.), *Vernunft, Handlung und Erfahrung, Über die Grundlagen und Ziele der Wissenschaften*, C. H. Beck'sche Verlagsbuchhandlung, Munich, 1981.

Translation copyright © 1987 by The University of Massachusetts Press
All rights reserved
Printed in the United States of America
Set in Linotron Sabon at G & S Typesetters
Printed by Cushing-Malloy and bound by John Dekker & Sons
Library of Congress Cataloging-in-Publication Data
Lorenzen, Paul.
Constructive philosophy.
1. Philosophy. 2. Logic. 3. Languages—Philosophy.
4. Mathematics—Philosophy. 5. Physics—Philosophy.
I. Title.
B29.L63213 1987   149   86-25003
ISBN 0-87023-564-8 (alk. paper)

This publication has been supported by the
National Endowment for the Humanities,
a federal agency which supports the study of such fields
as history, philosophy, literature, and language.

CONTENTS

Translator's Preface   vii
Introduction to the English Translation   ix

I. GENERAL PHILOSOPHY

1. Methodical Thinking   3
2. Enlightenment and Reason   30
3. Political Anthropology   42

II. CONSTRUCTIVE LOGIC

4. Protologic: A Contribution to the Foundation of Logic   59
5. Identity and Abstraction   71
6. Dialogical Foundation of Logical Calculi   78

III. PHILOSOPHY OF LANGUAGE

7. Logical Structures in Language   105
8. Logic and Grammar   113
9. Logic and Hermeneutics   121
10. Constructivism and Hermeneutics   130
11. Rational Grammar   135

IV. PHILOSOPHY OF MATHEMATICS: ARITHMETIC

12. How Is Philosophy of Mathematics Possible?   157
13. Justifying the Deductive Method   171
14. Pascal's Critique of the Axiomatic Method   175

V. PHILOSOPHY OF MATHEMATICS: ANALYSIS

15. The Actual–Infinite in Mathematics   195
16. Constructive Foundations of Mathematics   203
17. Classical Analysis as a Constructive Theory   208
18. Moral Arguments in Foundational Discussions of Mathematics   220

VI. PHILOSOPHY OF PHYSICS

19. How Is Objectivity in Physics Possible?	231
20. Constructive and Axiomatic Methods	238
21. Concerning a Definition of Probability	249
22. The Foundational Problem of Geometry	257
23. Geometry as the Measure–Theoretic A Priori of Physics	274

TRANSLATOR'S PREFACE

It is a scandal and also a very real danger that the twentieth century has produced so much scientific knowledge and so little understanding of the nature of scientific knowledge—its foundations, structure, and relation of the parts to the whole. Even philosophy, traditionally the home of the investigation of the nature of knowledge, has relinquished the task—preserving the rhetoric, but in reality seeking only to become yet another specialty discipline, erecting its fences and scrupulously respecting the fences of other departments of knowledge. Constructive philosophy, the work of Paul Lorenzen and his colleagues, is unique in the philosophical literature of the twentieth century in that it carries on the investigation of knowledge and demonstrates that it is not a vain endeavor.

The most noteworthy features of the constructivists' work are that it is both demonstrative and normative. In these essays Lorenzen demonstrates that scientific procedures and theories can be given a foundation in nonlinguistic procedures, thus escaping the infinite regress inherent in attempts to produce axiomatic foundations for science. He then demonstrates how, by using linguistic operations similarly founded, these procedures can be expressed in propositions and then further refined to construct individual scientific theories. Thus Lorenzen and the constructivists also illuminate the normative aspect of science, its ability to guide and regulate technical practice. In this way constructive philosophy also enunciates a regulatory principle for science, accepting as meaningful science only those procedures and propositions that can be demonstrated to be constructible from a foundation in nonlinguistic procedures and similarly founded linguistic operations.

The twenty-three essays by Lorenzen translated here represent collectively an overview of constructive philosophy and individually a series of demonstrations of the methodical foundation and structure of scientific knowledge. No further introduction is possible. Constructive philosophy is not a doctrine to be paraphrased or a theory to be justified by argument from general principles—the details to be filled in later. It is simply the methodical demonstration of the practical foundation and construction of knowledge contained in the following pages.

That this demonstration is now published and available in English is due to Richard Martin of the University of Massachusetts Press, who recognized the importance of the constructivists' work and who has seen the translation of these essays through to its completion with patience and perseverance.

KARL RICHARD PAVLOVIC
Gaithersburg, Maryland
July 16, 1986

INTRODUCTION TO THE ENGLISH TRANSLATION

First of all, my thanks to the translator, Karl Pavlovic. It was his idea to translate a collection of my essays (written in the last twenty-five years), and, as far as I can judge, he has succeeded in introducing the reader to the world of "constructive" philosophy.

Constructive philosophy is philosophy in the critical spirit of Kant, but with the tools of modern philosophy of science. The decisive points of Kant are maintained: the primacy of "practical" (i.e., ethicopolitical) reason over "theoretical" (i.e., technical) reason, justified by a demonstration that our theories in logic, mathematics, and physics are instruments made by us for our technical purposes. Parts II–VI give the essential steps of this demonstration.

For the reader familiar with modern philosophy of science it may be helpful to state that the term "constructive" (being the Latin equivalent of "synthetic") stands in opposition to "analytic." Both analytic and constructive philosophy are philosophies after the linguistic turn. They are the results of Frege, Russell, and Wittgenstein. Constructive philosophy is at the same time philosophy after the pragmatic turn (Peirce, Dingler, and the later Wittgenstein). In contrast to analytic philosophy, in constructive philosophy our talking about what we are doing (about our praxis) is *not* taken as something given that has to be analyzed but as something to be constructed by us. The construction has to be done methodically, that is, step by step, without gaps and without circles.

Methodical language construction starting from pretheoretic praxis and leading to theories supporting praxis: that is the program of constructive philosophy. Parts II–VI contain earlier essays (to provide more background) and essays that present the constructions in their latest form, especially chapter 6 for logical calculi, chapter 11 for rational grammar, chapter 17 for higher mathematics (so-called analysis), and chapter 23 for geometry, kinematics, and dynamics. Only for geometry is there a still later version in print, my book *Elementargeometrie,* (Mannheim, 1984).

When "science" in the sense of logic, mathematics, and physics has been *re*-constructed as the theoretical support for technical praxis, it has been demonstrated that science in this narrow sense gives us no orientation as to the ends for which we should use our technical means. Either we give up all attempts to find rational methods for arguing about ends, or we try to *re*-construct Kant's "practical" philosophy—though the term "practical" is misleading. After the pragmatic turn we know that we have to look for a pretheoretical praxis before any methodical language construction can begin. The praxis for which supporting political theories have to be constructed is the praxis of lawmaking, that is, the verbal praxis of arguing about legal norms.

For the practical engineer who decides what to do, theoretical knowledge (as mathematics and physical theories) serves as orientation for his decisions—but no theory makes the decisions. There is a corresponding relation between the practical politician and theoretical knowledge (as ethics and political theories). Political theory should serve as orientation for the decisions of practical politicians—but no theory should make the decisions.

Against the "analytic" suspicion that this correspondence is a misleading analogy, the essays of Part I lead in chapter 3 to a political anthropology that gives an orientation for lawmaking under the modern condition that no longer do traditional morals (or dogmatic religions or ideologies) provide such a generally accepted orientation. The moral principle of Kant is reconstructed for the lawmaker as the principle of rational argumentation, that is, argumentation that uses only impersonal, transsubjective arguments.

The ethicopolitical sciences (the "humanities") have the task of teaching the practical politicians (or, more generally, the politically interested citizens) the art of rational argumentation. We are back to logic and philosophy of language (including hermeneutics in chapters 9 and 10). Though the details of mathematics (Parts IV and V) and physics (Part VI) are irrelevant for political theories, Plato recommended geometry as the best preparation for rational argumentation. Later essays on political theory are to be found in my book *Grundbegriff technisches und politisches Kultur* (Frankfurt, 1985).

Finally, my thanks to the University of Massachusetts Press for taking the risk of publishing a philosophical book against the mainstream of analytic thought in the English-speaking world.

PAUL LORENZEN
Göttingen
December 1985

PART I
General Philosophy

ONE

Methodical Thinking

Motto: When we teach men *how* to think, instead of *what* to think, then we also reduce misunderstanding. It is a sort of baptism into the mysteries of humanity.
LICHTENBERG

In Plato's time there was as yet no controversy over whether mathematicians were justified in positing certain propositions at the very beginning of their investigations. Aristotle called these propositions axioms, as if, as Plato said, they needed no justification either from themselves or from some other source. It is said that Eudoxus tried to reduce all geometrical propositions to definitions alone. We can still find indications of this attempt in Euclid, but since Aristotle and Euclid the so-called axiomatic method has encountered no opposition and has been considered the only scientific method.

The axiomatic method can be formulated in two theses:

1. Human knowledge is based on certain undefinable fundamental concepts, and all further concepts are to be defined using the former.
2. Certain unprovable fundamental propositions concerning the fundamental concepts are true, and all further propositions are to be proved true using the former.

The Cartesian *Regulae ad directionem ingenii* repeat these theses in their essentials—and thus the science of the seventeenth century generally tried to add new axiomatic theories to the received Euclidean geometry. Mechanics was one such attempt, but in addition Spinoza and Hobbes tried to create axiomatic theories for ethics and politics.

Interestingly enough, Descartes's own geometry—so-called analytic geometry—exhibits tendencies that are not consistent with the axiomatic methodology. Analytic geometry reduced geometric problems to algebraic problems, that is, to the Indo-Arabic art of algorithmic calculation.

This algorithmic calculation, more generally arithmetic, does not rest upon axioms. At least that was the view at the time, for we do not find arithmetic axioms given. In any event, the foundations of arithmetic remained obscure and unilluminated. The only thing that was clear was that arithmetic was always a matter of operating with artificial symbols, which is something altogether different from the logical deduction of propositions expressed in words of a natural language.

At the time only Leibniz recognized the similarities between deduction and calculation and thereby anticipated the propositional calculi of modern logic.

In any event Plato's difficult question remained unanswered. How could there be a science that did not rest upon axioms, λόγον διδόναι? This question was treated seriously for the first time by Kant in his transcendental philosophy. But the fate of Kantian philosophy is known: Its laudable impulse ran far afield, particularly in Hegel, from the problem of securing a methodical basis for exact knowledge.

I believe that this explains why at the end of the nineteenth century the axiomatic method came back into revered favor, as if Kant had never lived, and why even now, thanks to the development of modern logic, it has such a hold on the claim to exclusivity that arithmetic is conceived as possible only as an axiomatic theory.

Logical positivism uses this axiomatic methodology as an instrument to proscribe as unscientific all philosophy and in general all intellectual inquiry within the sciences of man that fails to conform to this method.

If we examine the current state of the natural sciences, we find the same situation: Apparently, only axiomatic theories are possible. Considered without prejudice, however, the question of methods for thinking is not really a question for the natural sciences. If we do ask the question, however, it then falls within the purview of the sciences of man. The act of thinking is a human production; if we inquire into its method, we then need to *understand* this method as a human production too.

What then is the method by which we learn to understand? It looks at first as if the question has now shifted. Instead of considering the method of thinking, we are now dealing with the method of understanding. Instead of logic, we are dealing now with hermeneutics.

In the nineteenth century this shift led to a promising insight. I say promising, because with this insight it is possible to take up again the question of method in thinking, within which I now want to include understanding.

Hermeneutic thought from Schleiermacher to Dilthey issued in the following observation by Dilthey: "Knowledge cannot go beyond life." I have no intention of suggesting that an insight can be captured in a single sentence. Nonetheless, Dilthey's statement can well be taken as a formula that fixes a fundamental change in philosophy. This change, opened up in the nineteenth century and today, at least outside positivism, is beginning to manifest itself. Although Feuerbach's turn to the primacy of practice has had widespread political effect through Marx, it did nothing for theory of knowledge. Only Misch on the one hand and Heidegger on the other, both building on Dilthey and Husserl, have made clear what it means to say that thinking must begin with life, with the practical existence of man. All thinking is a refining stylization of that which has always constituted the practical life of men and women. From this perspective the philosopher no longer misunderstands himself as a consciousness that receives knowledge of the world only through perception, observation, and intellectual deduction—the usual misunderstanding since the time of Descartes and Locke. The world is seen as the unmediated present or ready-to-hand. I would say that philosophy has won a new immediacy.

That sounds very hopeful, but it is much too early to attribute much of a chance to this new beginning, which comes from the sciences of man. At the moment the kind of thinking that draws its examples from the natural sciences exercises a strong, perhaps still growing, influence, even in the sciences that deal with man.

And you will now skeptically inquire how philosophy then proposes to proceed from its apparently new immediacy. By what method will philosophy finally, once and for all, achieve some secure results?

It must be admitted that no hermeneutics that constitutes a communicable doctrine is possible unless it already recognizes and makes use of logical or, more generally, methodically ordered thinking. At this point in the discussion, humanists usually refer to the hermeneutic circle, that is, to the essential circularity of human understanding. They represent the search for a methodical beginning for thinking as a rationalistic illusion in which are caught, to their mind, only the naive natural scientists with their faith in progress.

Curiously, the above is not completely the case. Since the 1930s logical positivism, particularly in the cases of Tarski, Carnap, and Quine, has utilized a conception of thought and speech that ends in an inescapable circularity.

In logicistic philosophy the problem takes the form of asking for the foundations of scientific languages. The rules of logic are taken to constitute the syntactical rules of these languages. The answer to the question is represented most clearly by a simile in which a language is likened to a ship upon which we find ourselves under the restriction that we can never put in at a port. All repairs or improvements to the ship must be carried out on the high seas.

To be sure, this picture captures an essential truth, but it is used explicity by logicistic philosophy to block attempts to find a methodical beginning for thought. Every scientific language, which in order to be scientific must be representable as a calculus, has a semantics, that is, an interpretation of the symbols employed in the calculus. This semantics requires, however, a language that is called a metalanguage. In practice, this metalanguage is one of the natural languages. It is from this boat that, indeed, no one disembarks.

At this point hermeneutics and logicism coincide. Both schools of thought renounce the project of methodically reconstructing thought. To conclude from this coincidence that such a renunciation is necessary is to conclude from a mere fact. It seems to me much more appropriate that we be doubly cautious in the face of this coincidence.

I want to point out that Dilthey's proposition that knowledge cannot go beyond life does not prove that we must give up the search for a methodical beginning for thought (or knowledge). Dilthey's proposition says only that such a beginning is not to be found beyond life.

Someone will nonetheless object that life—the practical existence in which we always find ourselves before we begin to engage in science or even to philosophize—also includes the natural language that we speak. That is true, but it does not mean that we need use that natural language and its rules as the beginning of the methodical construction I am considering here.

If we envision natural language as a ship at sea, then our situation can be described as follows: If we are unable to make landfall, then our ship must have been constructed on the high seas—not by us but by our ancestors. Our ancestors must have been able to swim and have somehow carpentered together a raft out of, say, driftwood. They then continually improved on this raft until today the raft has become a comfortable ship. So comfortable that we no longer have the courage to jump into the water and once more start from scratch.

To solve the problem of the method for thought, we must put ourselves in such a shipless condition, that is, bereft of language, and then attempt to retrace the activities whereby we could, while swimming free in the middle of the sea of life, build for ourselves a raft or even a ship.

Because such an experiment is compatible with neither hermeneutics nor logicistic philosophy, I cannot draw upon any authoritative support. The only option that remains to me is to demonstrate here and now such a methodical construction.

I must make, however, one last preliminary observation. You might now suspect that I propose to leave off speaking English because I do not want to presuppose a metalanguage. That is not my intention. In what follows I will use the English language only to describe what you would have to do if you wanted to teach someone a language methodically. My descriptions could be replaced by practical instruction like that which is given to children when they are first learning to speak. Conversely, what I say here will only be substituting for such practical instruction.

If I occasionally remind you of certain commonplace observations concerning natural languages, this will be done only to bring you, who already have mastery of a natural language, more quickly into the picture.

I would like to begin with the stipulation that in all the natural languages that linguists have described to us (Chinese, Hopi, Ewe, or whichever) it is possible to combine sentences by syntactical means. I begin with this stipulation only so that we can examine sentences that are not syntactical combinations. Strictly speaking, this point of inquiry is meaningful only when applied to a specific language, but every language has equivalents to sentences of the simplest syntactical form: "This is so;" "This is not so." Sentences of this form can be called *fundamental propositions*. To be sure, such fundamental propositions are also constructions in which a subject is joined to a predicate. Therefore, they are themselves methodically preceded by sentences consisting of a single word in which the subject is replaced by the situation, an indicative gesture, or something similar. Only the predicate is spoken. It is always possible to introduce a new predicate by using a sufficient number of examples and counterexamples. I call this process the exemplary introduction of predicates. For example, which things are to be called "chair" and which are not, or when something is said to be "clean" and when not, are learned through exemplary introduction. I am not appealing here to facts of child psychology. This is not a matter of a scientific determination of the ontogeny, or even of the phylogeny, of speech. I am only reminding you that it is possible to introduce predicates exemplarily.

Now, it is obvious that the use of predicates that have been introduced only exemplarily remains very ambiguous. Nonetheless, exemplary introduction represents a possible beginning for speech.

When a predicate is used, it is always the case that a specific thing is referred to, to which the predicate is either attributed or denied. This specific thing does not need to have a proper name. For example, normally children learn the predicate "doll" before they give proper names to their dolls.

If we have proper names and predicates at our disposal, then we can put together fundamental propositions. Subjects are the proper names for individual things, and predicates are attributed or denied to the individual thing using a copula—in English, the words "is" and "is not."

Only propositions that concern individual things and that have *only one* predicate should be called fundamental propositions. A fundamental proposition may, however, have more than one subject as, for example, in the sentence "Max and Moritz are related."

The basic forms for fundamental propositions are therefore as follows:

affirmative fundamental propositions: $S_1, \ldots, S_n \; \varepsilon \; P$
negative fundamental propositions: $S_1, \ldots, S_n \; \varepsilon' \; P$

$S_1, \ldots, S_n$ are used here as variables for the subjects of the proposition, whereas P is used as a predicate variable.

In every natural language it is possible to predicate individual things in this way. Whether it is done with a copula, as in English, or without a copula is irrelevant for methodical thinking. Every predicate in a language represents a distinction; the specific thing to which the predicate is attributed is thereby distinguished from those individual things to which the predicate is denied.

Fundamental propositions, in which individual things are directly discussed, represent an area where thinking is well advised to follow natural language. Natural languages contain an abundance of distinctions, occasionally even a surfeit. One can say with Humboldt that only with language was our world given form.

Having first methodically fixed the usage of predicates by means of affirmative and negative fundamental propositions, we are by no means limited to thinking only about that which is objectively observable. When distinguishing among the properties and characteristics of our social reality—for example, customs and manners—we also need to begin methodically with examples. In the latter case, the person who learns the distinctions must actually participate in an activity as a meaningful totality, say, in a productive activity or in a game, in order to understand the differences. The person must have at least sympathetic understanding of the activity. In all cases, however, it is equally necessary that the person be acquainted *with the subject,* if he or she wants to understand the distinctions to be learned; all word usages that are simply idiosyncratic must be methodically excluded.

The predicates introduced exemplarily constitute a system of distinctions that can serve now as a basis for further distinctions. I call such collection of exemplarily introduced distinctions a *distinction base.*

It is then necessary to investigate how we could construct thought upon such a distinction base, that is, using fundamental propositions as our sole linguistic means. Natural languages have available a rich syntax (e.g., logical operators, prefixes, and suffixes for word construction).

If we are to proceed methodically, we must also create such tools. We must construct a rational syntax. In constructing a rational syntax we should preclude any appeals to natural language. In so doing, we will distinguish our intentions from the intentions behind the semantics presented by logicism, because the latter eventually appeals back to the syntax of natural languages.

As a first step that will lead us beyond fundamental propositions, we can restrict the ambiguity that attaches to the predicates of a distinction base. This ambiguity derives from the predicates' having been introduced exemplarily.

For this first step, which will lead us eventually to concepts, I must next give examples in order to be true to the method I am recommending here. Let us then assume that certain predicates that would be rendered in English by, say, "living being," "man," "animal," "plant," "raven" have been introduced exemplarily. Only by introducing *rules* can we specify more closely the usage of these predicates. You will immediately comprehend the rules I mean if I sketch for you figure 1.1.

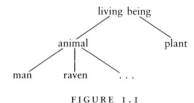

FIGURE 1.1

The rules for these predicates can be written as, for example,

$$\text{living being, } x \, \varepsilon \, \text{ animal} \Rightarrow x \, \varepsilon \, \text{ living being}$$
$$x \, \varepsilon \, \text{ plant} \Rightarrow x \, \varepsilon \, \text{ living being}$$
$$x \, \varepsilon \, \text{ animal} \Rightarrow x \, \varepsilon' \, \text{ plant}$$
$$x \, \varepsilon \, \text{ plant} \Rightarrow x \, \varepsilon' \, \text{ animal}$$
$$x \, \varepsilon \text{ living being, } x \, \varepsilon' \, \text{ animal} \Rightarrow x \, \varepsilon \, \text{ plant}$$
$$x \, \varepsilon \text{ living being, } x \, \varepsilon' \, \text{ plant} \Rightarrow x \, \varepsilon \, \text{ animal}$$

Rules like these are not universal propositions. Universal propositions cannot be given clear definition at this point in the development. Of greater practical interest at this point are directions prescribing how we move from one proposition to other propositions. The arrow symbol ($\Rightarrow$) is introduced here *exemplarily* to indicate such moves. The activity that is governed by these rules can be inculcated through practical training. The rules are neither commands nor sanctions, nor are they like the rules of a game. Such rules are much more the building blocks of a language to be learned, the building blocks of language games, as Wittgenstein said. If we read ⇒ as "if . . . , then," then this if/then is the practical if/then spoken of in logic.

We can, if we like, alter and reorder the use of the predicates given above in different ways, for example, so that man would no longer be considered an animal.

To search after the "true" system of such rules before the rules are given a practical employment would be in vain. We must begin with one set of rules or another before use and reflection can suggest possible improvements. Plato compared the situation with the art of cooking; you must first learn to butcher the carcass at the joints.

By way of an example of a system of rules for predicates found in use in the human sciences, I suggest figure 1.2.

FIGURE 1.2

Before you begin to discuss the concept "freedom" with another person, you should ascertain whether, for example, the other person accepts the distinctions implied by this system or whether—as is very often the case—the other person simply distinguishes under "action" only "agreeable actions" and "disagreeable actions" and understands "freedom" to be the same as "agreeable action."

Before you argue with another person about such systems of rules, it is important that you recognize that such arguments are arguments over rules.

The use of exemplarily introduced predicates can be more precisely specified by such rules. In introducing such rules, more and more predicates are encompassed by the rules. These further predicates are themselves more precisely specified by rules, and in this way systems of predicates are created. If, for example, we accept the specification given in figure 1.2 for the predicates "play" and "work," then we cannot at the same time call an action both play and work, even if the exemplary introduction of the predicates suggested such an equation. It is, however, always possible to change the systems of rules by introducing new examples.

The examples of rules governing the use of predicates that I have presented here produce what are sometimes called concept trees like those that result from the Platonic διαίρεσις. There are also other types of predicate systems (e.g., polarity systems). A simple example of the latter type of system is the black–white continuum in which a linear arrangement of the predicates for various shades of gray is given. More complex types of systems are required for so-called normative concepts like healthy, pure, just, and so on. We can arrange the predicates that indicate deviations from the norm in such a way that they all deviate to one side—for example, when justice is conceived as an ideal of which all real actions fall short. We could, however, also arrange the deviations so that the norm stands in the middle. Aristotle particularly used this last type for ethical virtues.

No matter how predicates are specificly bound together into a system by a specific set of rules, each predicate always acquires thereby a place value within the system in addition to the predicate's exemplary introduction. Within a given system the rules that constitute the system make it possible to carry out so-called deductions; for example, in the system given above the proposition "This is a living being" can be deduced from the proposition "This is a raven":

(1) $x \; \varepsilon$ raven
(2) $x \; \varepsilon$ animal
(3) $x \; \varepsilon$ living being

If the proposition $x \; \varepsilon \; Q$ is deducible from $x \; \varepsilon \; P$ in a given system, then we say that Q is deducible from P. Whether a proposition is deducible depends only upon the rules of the given system. It does not depend upon the nature of the exemplary introductions of the relevant predicates. We can now abstract from the level of exemplary usage. In complicated systems it sometimes happens that Q is deducible from P and, the opposite, that P is also deducible from Q. P and Q are then said to be *equivalent* in that system. In such cases we also say that the predicates P and Q "represent one and the same concept." In this way talk that deals with concepts is methodically introduced. This process is called *abstraction*. Concepts are often introduced as the *meanings* of predicates, in which case meaning has been conceived on analogy with the referents of proper names. A proper name refers to or means an object—there is nothing problematic about this, for this is precisely the function of proper names. Whether predicates also have meanings has been much debated in modern logic; even the old battle between realism and nominalism has been exhumed once again. Talk of concepts is not a problem, however, when concepts are introduced as I have done here. We have just indicated what we mean when we say that two predicates represent the same concept. We must now determine how to talk intelligibly about concepts themselves. To do that we must indicate how we advance to predicates that concern concepts.

We make this advance by first introducing predicates concerning predicates, that is, predicate predicates. We can, for example, exemplarily introduce a distinction between long predicates and short predicates. We do this by placing the predicate to be discussed in quotation marks and obtain, for example, the following propositions:

"long" ε short
"predicate predicate" ε long
"short" ε' long

We now restrict our attention to those predicate predicates that, when they are valid for a predicate P are also valid for every equivalent predicate Q. Such predicate predicates may be called *invariant*. The two-valued predicate predicate "is deducible from" is an example of an invariant predicate predicate.

To speak of concepts is to abstract from all those features that distinguish two or more equivalent predicates from each other. That is, to speak of concepts is to restrict oneself to invariant predicate predicates. When R is an invariant predicate, instead of "P" ε $\check{R}$ we will write $/P/$ ε R and read this as: The concept P is R. The use of the word "concept" indicates that an invariant predicate predicate is being attributed to P.

This restriction to invariant propositions constitutes the essence of the act of abstraction.

The act of abstraction transforms a given system of predicates into a conceptual system. The constituting rules of a conceptual system can therefore be called *conceptual specifications*. Conceptual systems can be viewed as the first rafts that we build as we swim upon the high seas unable to reach land.

The theory of concepts that I have sketched is not a theory concerning existing things; it is not an ontology. Concepts as described here belong to our actions; they are interpreted here functionally rather than ontologically. Moreover, we should not confound the theory of concepts presented here with logic, as did Aristotle. Logic, that is, the theory of logical operators, represents a further step that has not yet been taken.

How can logical operators be introduced methodically into the language that we have constructed so far? The starting point for this is found by returning again to the practical situations in which we speak. We must imagine two people, both of whom use the same conceptual system, perhaps the system above, and who engage in dialogue with each other. What does it mean when one of the two asserts, for example, that "all ravens are living beings"? This sort of assertion is not at all like any of the rules belonging to the conceptual system. In Aristotelian logic this proposition, "All ravens are living beings," would be interpreted as a logical relation between the concept "raven" and the concept "living being." Remarkable as it may seen, it was only in modern logic—thanks mostly to the investigations of Frege in his *Begriffschrift* of 1879—that it became clear that such propositions are constructed using fundamental propositions with the help of logical operators. In fact, such propositions are so constructed that we can reformulate them in English as

For all x, if $x \; \varepsilon$ raven, then $x \; \varepsilon$ living being

This can be formulated also symbolically as

$\bigwedge x \; (x \; \varepsilon \; \text{raven} \rightarrow x \; \varepsilon \; \text{living being})$
$\bigwedge x \; (x \; \varepsilon \; R \rightarrow x \; \varepsilon \; L)$

In this case two logical operators are used in the construction: the conjunction if/then and the quantifier $\bigwedge$, "for all." We have not yet determined the sense of these two operators. For that we must first indicate how such constructed propositions are used in dialogue.

If you assert a proposition, that means you are willing to defend the proposition against the attacks of a partner in a dialogue, against an opponent. For such acts of assertion and defense to be possible in general, we must first indicate how the relevant logical operators are to be used.

The normal usage of "all" in English suggests the following rule for the universal quantifier $\bigwedge$: The opponent may choose an x, say, "Hans," and then the first speaker, the proponent of the proposition, must defend the new proposition that results, namely, "Hans ε raven $\rightarrow$ Hans ε living being." To proceed with the defense of this proposition, we must indicate how the logical conjunction $\rightarrow$ is to be used. This can be done as follows with the help of the practical if/then.

If the opponent asserts the if-proposition "Hans ε raven" and can defend this assertion, the proponent must then assert the then-proposition "Hans ε living being."

A circular definition or infinite regress from metalanguage to metalanguage is avoided by using the practical if/then to explicate the logical if/then.

In the case of the assertion of primitive propositions like "Hans ε raven" and "Hans ε living being" we can assume that the partners in the dialogue agree either that the assertion is correctly made or that it is not. They agree either that it is true or that it is not, as we also sometimes say.

In this particular case, in fact, the proponent will win the dialogue and thus be in the right, completely independently of who this Hans is, that is, irrespective of which assertions concerning Hans are correct. Because the opponent asserted "Hans ε raven," the proponent can defend his assertion "Hans ε living being" merely by appeal to the relevant conceptual specifications. According to these conceptual specifications, the proposition "Hans ε living being" can be deduced from the proposition "Hans ε raven." We will say that this deduction is a *proof* of the assertion. The proof in this case uses only the necessary conceptual specifications, as opposed to any acquaintance with individual objects such as Hans.

Some propositions can be so defended that we do not even need to refer back to conceptual specifications. "All ravens are ravens" is a trivial example. More generally, this is the case with every proposition of the form

$$\bigwedge x \, (x \, \varepsilon \, P \rightarrow x \, \varepsilon \, P)$$

The dialogue runs as follows:

Opponent	Proponent
	$\bigwedge x \, (x \, \varepsilon \, P \rightarrow x \, \varepsilon \, P)$
? s	$s \, \varepsilon \, P \rightarrow s \, \varepsilon \, P$
$s \, \varepsilon \, P$	$s \, \varepsilon \, P$

A proposition that can be defended *on the basis of form alone* is called a logically true proposition. By the *form* of a proposition we mean the type and manner of its construction using logical operators. I want now to give a further example which also uses the additional logical conjunctions $\vee$ = or, $\dot\wedge$ = and, $\neg$ = not. Every proposition of the form "$a \vee b \dot\wedge \neg a \rightarrow b$ (*if a or b and not−a, then b*) is logically true. This logical truth results *not* because propositions of this form sound obvious to everyone who speaks English but rather because these propositions, in fact, always win in dialogue. The dialogue runs as follows:

	Opponent		Proponent	
(1)			$a \vee b \dot\wedge \neg a \rightarrow b$	
(2)	$a \vee b \dot\wedge \neg a$		L ?	
(3)	$a \vee b$		?	
(4)	a	b	R ?	b
(5)	$\neg a$		a	

In line 2 the opponent must assert the if-clause of "if $a \vee b \dot{\wedge} \neg a$, then b," that is, $a \vee b \dot{\wedge} \neg a$. As indicated by the dot over $\wedge$, this clause is a conjunction of $a \vee b$ and $\neg a$. The proponent now may question the left (L?) and the right (R?) sides of the conjunction. In line 2 he asks about the left side, $a \vee b$. In line 3 the opponent asserts the left side. The proponent questions this assertion (?), and the opponent may now (in accordance with the sense of $\vee$) himself choose whether he wants to defend the left side, a, or the right side, b, of $a \vee b$. There are thus two cases to consider. In the first case the opponent chooses in line 4 to defend a. The proponent then questions the right side of $a \vee b \dot{\wedge} \neg a$, and the opponent must answer in line 5 with $\neg a$. The proponent can contradict this by asserting a, which the opponent has himself asserted. In the first case the proponent wins. In the second case the opponent chooses b in line 4. Here we have a simple proof of a logical truth. The individual steps of the proof are based not on axioms but rather on the definitions of the relevant logical operators, that is, their use in dialogue as defined by the rules.

Even though these examples are quite simple and the logical operators are easily compassed (there are exactly six operators: $\dot{\wedge}, \vee, \rightarrow, \neg, \wedge, \vee$), the iterative conjunction of these six operators permits the construction of forms of whatever complexity one desires. It has already been proven that no calculating device can be built that can answer every possible question about the logical truth of formulas. That is a mathematically very interesting result. In fact, logicism displays this result proudly as one of the showpieces of its new philosophy.

Once we have the logical operators at our disposal, the usual kind of definition, which I want to call more precisely *analytic definitions*, can be used. For example, we can define "sorrel" using

$$x \; \varepsilon \; \text{sorrel} \Leftrightarrow x \; \varepsilon \; \text{horse} \wedge x \; \varepsilon \; \text{brown}.$$

Such analytic definitions are basically a convenience and can be dispensed with; they only abbreviate a logically complex expression. When the predicate to be defined has been introduced exemplarily prior to being defined, the definition is a new kind of conceptual specification. Although in natural languages we never know whether a given partner in a dialogue associates certain exemplars, rules, or definitions with a predicate, the point of constructing a language methodically is precisely to certify that over which there can be no doubt.

Instead of going on into the problems of formal logic that arise at this point, I would rather sketch out our further construction of this first boat made out of conceptual systems and formal logic, a boat in which we are already sailing somewhat more securely upon the sea of life.

As I indicated earlier, our efforts concerning methodical thinking are undertaken within a specific context. Both hermeneutics and logicistic philosophy explicitly exclude the possibility of a methodical beginning for critical thought. The reason for this is found, inter alia, in the Diltheyan principle that knowledge cannot go behind life. If we take "life" to include the possession of a natural language that is an accomplished and given fact for each person, then this exclusion appears to be necessary. Without going behind life, however, we can nonetheless separate what we previously spoke from that which we now want to begin to speak methodically and critically. Thus we can attempt to understand the logic that we are given while we methodically reconstruct it; such a project is the program of a hermeneutic logic. At the same time, the language of hermeneutics must be methodically acquired; that project is the program of a logical hermeneutics.

If we compare our situation in life with that of a group of people shipwrecked, then our task is to build ourselves a ship in the middle of the ocean. The first planks are predicates, which we first fashion using distinctions offered us by happenstance. With rules we then fasten these predicates together into conceptual systems. When we next add the logical operators, the sense of which is then fixed within the context of dialogue, to the conceptual systems, it seems to me that we have already done a good piece of work. To be sure, it is a long path from this point on through many steps to the prose masterpieces of the human sciences or to the formulas of a physical theory. Some surprises may also await us along the path, but at least a beginning has been made.

The syntax of our natural languages is immensely richer than logic and its six operators. Some of that syntax is at present dealt with in formal logic, for example, the usage of definite and indefinite articles. But a rational syntax is also required, for fundamental propositions need not belong to a formalized language. On the contrary, fundamental propositions can be constructed using predicates specified only exemplarily.

Other aspects of natural syntax need not be reconstructed in a rational syntax, because they can be replaced by adverbs and then can be reconstructed as higher-order predicates. This is the case, for example, with the syntax of temporal words, that is, words that contain time specifications.

Using very simple examples we can make the essentials clear. When, say, a bird sings and we say "The bird sings," we are not dealing with a fundamental proposition because the subject of this proposition is not a proper noun. This is a case of two predicates. The first predicate serves to indicate the object to which the second predicate is attributed. In English such cases are indicated by the use of the definite article; in Latin the article is missing, *avis cantat*. The sense of this sort of construction becomes clear when, for example, we reword the proposition as follows: "That, which is a bird, sings." When we put it into a formal symbolism, we get: $\vee\, x\, (x\, \varepsilon\, P)\, \varepsilon\, Q$. For the methodical introduction—that is, without appeal to English syntax—of propositions of this form, we must imagine how, for example, children go from the use of one-word sentences like "bird" and "sings" to the complex sentence "bird sings." In order for the first predicate to be able to replace a proper noun it must be drawn from the context in which predicates are used to refer to objects univocally.

When we add yet another predicate to our example, say, "beautiful," further possibilities arise—for example, the possibilities that are instantiated in English by:

> The beautiful bird sings.
> That the bird sings is beautiful.
> The bird sings beautifully.

The first example is a case of logical conjunction in which the object is characterized by its being a bird *and* its being beautiful. In contrast, in the second example "beautiful" is a predicate attributed to a fact. Facts in this sense result from an act of abstraction from a proposition that corresponds to the abstraction of concepts from predicates. Such a predication of a fact occurs, for example, if you say in English "The bird really sings." In this case it only looks as if we have here an "adjective" that says something about singing.

It is clear that, once a rational syntax begins to be used, it can be extended further. The governing principle is that we never make use of an existing language in methodically extending the syntax. To be sure, we can always—as in the metalanguages of semantics—speak about the constructed language, but the linguistic mechanics of this speaking-about-speaking must be introduced using the same methods as those used to introduce the mechanics of the basic language. Only the subject of the metalanguage is different; in the metalanguage we draw distinctions, for example, not between men and animals but rather between subjects and predicates. The metalanguage is the language in which we reflect upon and consider the basic language. The syntactical resources of the metalanguage must never anticipate the resources of the basic language.

The importance of the act of constructing a rational syntax lies in the fact that only by doing this do we come to understand our own thinking and speaking. The principle that we only understand that which we could ourselves produce leads us—if we inquire into its historical origins—into regions far removed from hermeneutics. This principle played a large role in the work of Vico. It is also, however, a decisive theme in Kant's transcendental philosophy. He once said, "We understand nothing so well as that which we could also make, if we were given the necessary material."

This will to understand correctly was the most important driving force behind the natural sciences in the period from Descartes, through Kant, up to the close of the last century. Although we can trace the beginnings back to the so-called discovery of non-Euclidean geometries and the constitution of electrodynamics as the first part of physics not reducible to mechanics, it was modern physics that first brought forward the positivistic renunciation of understanding. Things are only to be "described." There are also people who are even content with "empirical confirmation" of rules that they do not understand, that no longer even claim to describe something.

In any event, although it may not trouble some physicists to renounce understanding, hermeneutics must nonetheless try to understand physicists. Physics is produced by human beings.

I would rather not go into any physics at this point, but I must make clear and understandable the conditions under which an exact science is possible. I must do this if I am to refute the belief that this business of finding a methodical beginning for thought in practical life is at most relevant only for the human sciences.

First, we need to understand pure mathematics. Pure mathematics means arithmetic in all its branches—differential calculus, the theory of functions, and so on, but excluding geometry.

In modern textbooks arithmetic is developed axiomatically. The origin and descent of the axioms used remain in the dark. These axioms were, as I indicated, first formulated at the end of the last century. Their formulation was really superfluous, because Kant had previously clearly indicated whence our exact knowledge concerning numbers came—namely, from the fact that we ourselves construct these numbers.

We can exemplarily introduce individual number words and can then add conceptual specifications or analytic definitions, but we will never arrive at an infinite series of numbers in this way. To obtain an infinite series we need a construction. The simplest form of such a construction begins with a number sign, for example, /, and then proceeds according to a rule that states that to every sign n a further line is to be added, namely, $n \Rightarrow n/$. To be sure, it is not in practice possible to produce an arbitrary quantity of signs by following this rule (e.g., 10 to the 100th power), but that is only because our lives are too short, the supply of chalk is too small, or something similar. According to this rule any given quantity of signs is, as they say, *theoretically* possible. The meaning of the phrase "theoretically possible," if it is unfamiliar to you, must be learned from examples such as this one.

The construction formula

$$\Rightarrow /$$
$$n \Rightarrow n/$$

supplies what I call, in a loose allusion to Kant, a *synthetic definition* of the concept of number. There is in addition an act of abstraction that leads from number signs to numbers themselves, but it is not important here. Using synthetic definitions the various forms of calculation can be introduced. We can define, for example, how addition and multiplication are to be performed:

$$\Rightarrow \frac{m + /}{m/} \qquad \Rightarrow \frac{/ \times n}{n}$$

$$\frac{m + n}{p} \Rightarrow \frac{m + n/}{p} \qquad \frac{m \times n}{p} \quad ,, \quad \frac{p + n}{q} \Rightarrow \frac{m/ \times n}{q}$$

Arithmetical propositions are logical consequences of these definitions—propositions that can be defended against anyone in dialogue using these rules. Arithmetical propositions are not analytic propositions, because they are not deduced from analytic definitions. They are synthetic propositions. In addition, they are synthetic propositions a priori, because they cannot be refuted by experience. In fact, such synthetic definitions are what make the counting of real things possible.

In higher mathematics, so-called analysis, the situation is no different in both essence and method. It is only that more complicated construction formulas are added. None of this can be discovered in mathematics textbooks; there you will find out only that mathematics is an axiomatic science. Faith in axiomatic mathematics is the accepted solution to the foundational crisis that came to a head around 1900 with the so-called contradictions in set theory.

This foundational crisis has a long history. It began with the Pythagorean discovery of incommensurable lengths; received its first solution from Eudoxus (recorded in the fifth book of Euclid's *Elements*); and broke out with new severity once again during the seventeenth century with the development of modern mathematics of infinitesimals. The successful applications of modern mathematics at first suppressed the problem of a rigorous foundation. It has only been since the nineteenth century that there has arisen a new sort of person, the pure mathematician who has no interest in applications. With the pure mathematician there also arose foundational research in mathematics. Foundational research tried to become clear about what had really been created with modern mathematics.

The most obvious difference from ancient mathematics was the thoroughgoing use of symbols in modern mathematics. The point of modern mathematics was the construction of symbols. The Greeks had used only the cardinal numbers 1, 2, 3, . . . , and also, at most, rational numbers like $2/3$, $3/5$, . . . , so that with decimal fractions we had something really weird, namely, something infinite. A decimal fraction is nothing other than a specific infinite series of rational numbers. Instead of infinite series we could also consider infinite sets. For the sake of simplicity I will restrict myself here to sets of cardinal numbers. The essential question is, "What is a set?" I offer a heretical answer that follows Weyl.

If we have arithmetic formulas with a free variable x for cardinal numbers—for example, $x > 2$ or $x + 1 > 3$, then it can be the case, as it is here, that two different formulas $A(x)$ and $B(x)$ are true for the same numbers. That of course means only that $A(x) \leftrightarrow B(x)$ is arithmetically valid. We say in such cases that the formulas "represent the same set" and write

$$\varepsilon_x A(x) = \varepsilon_x B(x)$$

The transition from formulas to sets is an act of abstraction that can be carried out utilizing the same procedure that we encountered earlier in the case of concepts.

For numbers and sets the elementary relation ε is defined by

$$y \,\varepsilon\, \varepsilon_x A(x) \Leftrightarrow A(y)$$

All sets are now represented by formulas—what kinds of sets exist depends on what kinds of formulas are available. In arithmetic, primary formulas can be introduced using synthetic definitions, as in the cases of addition and multiplication. Further formulas can be constructed logically using the primary formulas. The resulting sets form, as I say, a level of sets over the cardinal numbers. The size of this higher level depends on our construction of formulas, so that we can always expand it.

In contrast to this constructive set theory, axiomatic set theories run as follows. In addition to number variables, $x, y, \ldots$, axiomatic theories take an additional kind of variable, $M, N, \ldots$, that are called set variables and that add to arithmetic a new kind of formula, $x \,\varepsilon\, M$. Using arithmetic formulas, the axiomatic set theories then logically construct additional formulas in which there appear not only number quantifiers $\bigwedge_x, \bigvee_x$ but also set quantifiers $\bigwedge_M, \bigvee_M$. When $A(x)$ is such a formula that also contains a free number variable x, the following comprehension axiom is needed:

$$\bigvee_M \bigwedge_x . . x \; \varepsilon \; M \leftrightarrow A(x).$$

This axiom appears to be harmless. If we interpret M as a variable for the sets that appear in a given level of sets above the cardinal numbers (and if no set quantifier is added to $A(x)$), then $\varepsilon_x A(x)$ really belongs to that level. The axiom above requires, however, that the level be so constructed that a set M with $x \; \varepsilon \; M \leftrightarrow A(X)$ also be in that level if there are any set quantifiers in the formula $A(X)$. No one has yet succeeded in constructing such a set. Therefore, although in arithmetic we can justify axioms by appeal to a constructive theory, we cannot justify this axiom of axiomatic set theory by such an appeal. In my opinion this means that axiomatic set theories are simply a fantasy product.

What is going on here is that people are proposing axiomatic theories that use signs like $x \; \varepsilon \; M$ as formula symbols, where ε is uninterpreted. Because we can propose as many axiomatic theories as we like, the only objection to this is that the significance and purpose of this activity remain completely obscure. On the other hand, if in the future someone were able to construct levels of sets for which the axioms are true, then an axiomatic foundation for set theory would be superfluous.

What are we to make of the fact that axiomatic set theories are recognized in mathematics as the only acceptable theories, despite the obvious unintelligibility of such theories? When we ask this hermeneutic question, we unfortunately reach into a wasp's nest of modern scientific prejudice. It is said that these theories are aesthetically pleasing or that they are empirically confirmed. These are the more harmless answers to the question. From more aggressive quarters comes the counterquestion: Do you doubt the validity of modern physics, which as everyone knows is built upon the foundation of the mathematics of infinitesimals?

It is certainly true that physics is built upon the foundation of modern mathematics, but no one has yet demonstrated that physics would not flourish *even better* with an intelligible (constructive) mathematics of infinitesimals than it has so far with an unintelligible (axiomatic) mathematics. None of the dogmatic empiricists has yet claimed that the axioms of set theory are the result of empirical observation.

The question of the connection between mathematics and physics makes an early appearance in geometry, particularly in the so-called continuum problem. Changes in place represent the obvious prototypical continuums—time and space also appear to be continuous.

The most remarkable fact about the continuum problem is that it was solved over 2,000 years ago. The solution is in Book 6 of Aristotle's *Physics*—but apparently has gone unnoticed. Aristotle defines a continuum (συνεχές) as διαιρετόν εἴς ἀεί διαιρετά, as an ever further divisible division. According to this definition the parts must always be themselves continuums; that is, the parts, taken as a whole, should be indistinguishable. According to this definition, there are, for example, on any length any number of possible points of division—whether we construct the division points as rational points or as end points of series of rational points is irrelevant.

According to set theorists, a given length is supposed to be a set of points, so that a further point can be assigned to every partial set, say, as a subordinate boundary. From this perspective the comprehension axiom above contains conditions for the existence of points; the set theorists believe that they have thereby "described" the continuity of a given length. As Weyl once observed, "that is all . . . obvious nonsense."

In my view the difficulty lies largely in the fact that there is a great deal of confusion over what in general constitutes a theory of spatiality, that is, a geometry. Beginning with Euclid, geometry has been presented as an axiomatic theory, not as a constructive theory. Indeed, geometry is not in its essence a constructive theory. In geometry we do not operate with symbols that we have produced, as is the case in logic, arithmetic, and set theory.

We can, of course, construe any axiomatic theory as purely a formalism. Although for certain kinds of investigations this is an appropriate procedure, it is not at all appropriate in any attempt to understand geometry as a human production.

Modern physics construes geometry to be a physical theory, an empirical theory of the spatiality of rigid bodies. The difficulty with this construal lies in defining the concept of a rigid body. To determine whether or not a body is rigid you need certain physical data, if you do not have geometry available. But to acquire these physical data you need to be able to take measurements, specifically geometrical measurements. To take geometrical measurements you need a rigid body. We can escape from this circle if we avoid construing the geometry received from the Greeks as either a purely formal theory or as an empirical physical theory. Rather, speaking as Kant did, we need to understand geometry as a theory about the conditions under which spatial measurements are possible.

One of these conditions is that we be able to check the rigidity of "rigid bodies used for measuring" (rulers, yardsticks, compasses, etc.). We can do this by paying attention only to the form of the body in question. All physical bodies are bounded surfaces, which are bounded by curves, which in turn are composed of points. We will never find among real surfaces one that is perfectly flat. The concept of flatness presents us with the formation of a new kind of concept. If we were to begin by introducing flatness exemplarily and then to proceed to conceptual specifications, we still would not understand why it is that ever more precise techniques of measurement produce ever more precise verifications of geometrical propositions. This indicates that flatness is a kind of concept different from the concepts we have considered previously. As is indicated by phrases like "absolute flatness," flatness is what Plato called an idea. We define a flat surface by placing a restriction on it; all of its points are to be indistinguishable. As was first articulated by Leibniz, this indistinguishability means homogeneity. Everything that is true of one point on the flat surface is to be true of all others points on the surface. To put this more clearly and precisely, if a proposition $A(E,P)$ is valid over a surface E and for a point P on E, then the proposition $A(E,P')$ is valid for every other point P' on E. Thus, for flat surfaces we need a principle of homogeneity:

$$P \; \varepsilon \; E \wedge P' \; \varepsilon \; E \wedge A(E,P) \rightarrow A(E,P')$$

Although this homogeneity does not by itself define a flat surface, the homogeneity principle does give us the first proposition of a synthetic–a priori geometry. We say synthetic because it is not built on analytic definitions; we say a priori because a measuring rod, a necessary condition of the possibility of measurement, is first defined with the help of the idea of a flat surface. Finally, we say geometry, not a purely formal theory, because this homogeneity concerns the boundary surfaces of real physical bodies. As Dingler first recognized, in practice flat surfaces are produced in accordance with this idea of homogeneity, for example the three-plate method of grinding flat surfaces.

If you use only two plates for grinding, spherical surfaces are produced. The indistinguishability of both sides of the surface is also part of the idea of a flat surface.

Using homogeneity principles, we can define not only flat surfaces but also other spatial forms, for example, orthogonality and parallelism.

Anyone familiar with geometrical instruments (e.g., a drawing square) knows what a right angle is supposed to be (viz., indistinguishable from its adjacent angle) and what parallel lines are supposed to be (viz., all points of both lines are to be indistinguishable). As in the case of flat surfaces, the forms of right angles and of parallel pairs of straight lines can also be specified using homogeneity principles.

These homogeneity principles constitute the beginnings of a systematic construction of geometry. They can also be construed as axioms of a theory, if we overlook the fact that the principles are restrictions placed on the points, lines, and surfaces of real physical bodies.

It can easily be demonstrated that the three basic forms mentioned above, flatness, orthogonality, and parallelism, suffice to provide a foundation for all of geometry. Using these forms we can define when two lengths can be said to be of equal length. Following that, we need to define parallelograms, namely, four-sided figures in which opposite sides are of equal length. In the special case of a rhombus, defined as having orthogonal diagonals, all sides are of equal length. With the parallelogram and rhombus we can check any set of lengths to determine whether they are equal (see Fig. 1.3). With that we can now define a rigid body: Equal

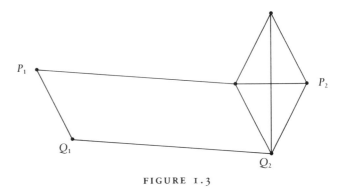

FIGURE 1.3

lengths marked on the body by two points must remain equal through all movements of the body.

With a rigid body we can finally define measurement of length. If, for example, we produce a most precisely flat surface containing a right angle and mark off three units and four units along the arms of the angle (as in Fig. 1.4), we will then find that the length of the hypotenuse is five units. As far as exactness of measurement is concerned, we can know this *beforehand* (which is simply the sense of a priori), because this length of five units is logically deducible from the homogeneity principles, the definitions utilized, and the assumptions made.

According to the empiricist conception of geometry such a priori statements are not possible. In any event, the logical possibility of non-Euclidean geometry and, more generally, the mathematical theory of so-called Riemannian warped space do not represent an argument against an a priori geometry. These prove only that *logically* other standards of length measurement are possible. In relativity theory we have in fact physically realized a new standard of length measurement, namely, the

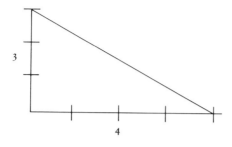

FIGURE 1.4

use of light signals. In precisely this case, if you will attend closely to the methodical ordering of theories, you will see that relativity theory presupposes the electromagnetic theory of light. The latter theory in turn presupposes mechanics, because mechanics makes it possible in general to measure forces. But mechanics in turn requires geometry.

So here too there appears to be a circle if we represent relativistic geometry as the only geometry that is physically meaningful. This circle progresses in a methodical order, however, if we begin with Euclidean geometry as the theory of homogeneous spatial forms and then construe Einsteinian relativity theory as a very much later theory with a different region of application (e.g., the optical measurement of distances). Relativity theory is then seen to be a physical extension of Euclidean theory rather than a refutation. These methodological questions are somewhat complicated by the fact that there are many intermediary steps between Euclid and Einstein. The two most important of these steps are chronometry (i.e., the theory of time measurement) and mechanics (i.e., the theory of mass measurement).

Time—it seems—is an inexhaustible topic of conversation in philosophy. But the relation between existence and time is not the subject of this discussion. The problem at hand is rather the measurement of time—that is, what can we say theoretically about the watchmakers' craft?

It would be pleasant at this point to quote Saint Augustine who said he knew exactly what time was but only when no one asked him about it. I think, however, it would be more instructive to quote Aristotle, who had a concrete answer to the question: ὁ χρόνος ἀριθμός ἐστιν κινήσεως κατά τὸ πρότερον καὶ τὸ ὕστερον, "Time is the number of movement with respect to the earlier and the later."

This answer is correct, because it assumes only that we are already acquainted with movements and can already distinguish earlier and later within movements; these predicates are in fact easily introduced exemplarily. Time is defined as a measure for movements, but, unfortunately, we are not told how the measurement numbers are to be obtained. The Aristotelian specification requires such a supplementation. We need to determine when two occurrences of movement have equal duration. This was no problem for Aristotle, because he took the fixed stars as the archetype of uniform movement, that is, an ideal clock. Today we explain the "movement of the heavens" as the result of the earth's rotation, as did the Pythagoreans. But today we also have reason to believe that the earth's rotation is not uniform due to, for example, the braking effect of the tides. Of course, that is only possible when we have reason to believe that some other movement is uniform. There is buried in this an a priori principle, or a synthetic definition as I prefer to call it. In the case of a repetitive process, for example, a swinging pendulum, we call the duration of the individual actions equally long only if the individual actions are indistinguishable.

Unfortunately, when it comes to the basic concepts of mechanics Aristotle can no longer provide us with details. The founders of modern mechanics from Galileo to Newton have left us no usable definitions. As a consequence, the foundations of mechanics are still today a morass that physicists are well advised to avoid. The mathematical theories involved in mechanics, and indeed also classical mechanics itself, are unobjectionable as mathematical theories. We know how to apply these theories in practice, but we are lacking any theoretical insight into what we are really doing when we apply them.

I think that Weyl provides the best suggestion for achieving some clarity here. He suggests that the law of conservation of momentum be taken as the definition of the inertial mass of physical bodies. If two bodies collide, say, inelastically, so that they remain in contact after the collision, and if the speeds of two bodies are v_1 and v_2 (relative to the system of the bodies after the collision), then for their masses $m_1 v_1 = m_2 v_2$. It is probable that historically this definition of mass played an important, if not explicit, role in the work of Galileo, Descartes, and Huygens. Otherwise, it is hard to understand why the conservation of momentum is given the same a priori necessity as the propositions of geometry.

With the concept of inertial mass Newton was able to define force using the product of mass and effective acceleration. From only these definitions, to which was added only one natural law (the Newtonian law of gravitation), resulted all of classical modern mechanics.

I will not at this point take up the concept of causality, which comes not from physics but from practical life. Causality never appears in the formulas of physics. The course of the transformations of a mechanical system is clearly determined by the system's initial conditions.

Even in Kant's time there was the suspicion that all of mechanics—and that meant then all of physics—must be, like geometry, deducible from fundamental synthetic definitions. Today, on the contrary, it is clear that the constitution of the basic concepts is only a preparatory step, only after which does what we call physics begin (e.g., electrodynamics, relativity theory, and quantum mechanics). Geometry and rational mechanics do not belong to physics in the modern sense. Rather, these two represent a preliminary stage, a *protophysics,* as I like to call it.

Whether empirical physical theories can lead to a revision of protophysics, as the question is today conceived, is a question that typically belongs to foundational research in physics.

To understand that this question is of fundamental consequence for the method of physics, it is not enough to study physics as it exists today. This is a question that normally lies outside the interest of physicists. The physicist will most likely view this question as a "philosophical" question, although there are exceptions. This view rests upon the division of labor that has become usual in the modern business of science. In reality, however, reflection upon what one does when one engages in science is not separable from science.

Please permit me, in closing, to go into the problem that has arisen here: the problem of the distinction between science and philosophy. What does reflection on the possibilities of methodical thought yield with regard to this distinction, a distinction that is today so open to question?

Received philosophy is a truly complex phenomenon. If, however, as a nucleus we take from it only that part which at least in intention confines itself to propositions that can be secured methodically, then within this nucleus of critical thought there is no boundary between science and philosophy. It is indeed a fact that reflection on methods is frequently not well developed in the individual sciences, but as a fact this signifies nothing. It belongs to the current style of science. In any event, the style of ancient science did not lack reflection.

I want to point out that perhaps none of what I have said here about methodical thought is beyond any scientist, from rational syntax to protophysics. It is true that reflection, when extended to that which is common to all special sciences (e.g., logic), no longer falls within any one science. That is obviously the case. But reading and writing also belong to every science and are nonetheless not philosophical.

A theory concerning the method of thought can be counted as an essential part of the propaedeutic of science, as elementary instruction, if you like. This was the case when the arts faculty was still called the lower faculty—and did not feel itself particularly aggrieved thereby.

But, because the natural sciences on the one hand and the historical sciences on the other have emerged from this arts faculty, there has no longer been an intelligible place for critical philosophy that is divorced from science. And when we demand of both science and philosophy that they proceed methodically, philosophy still has no place of its own. Philosophy cannot be denominated critical foundational research and be separate from the sciences. Nor can philosophy be something that comes only after science, as a crowning dome placed upon the scientific edifice. A philosophy that makes itself dependent upon the dogmas of science must assume an uncritical stance toward science—and thereby destroy itself.

There remains, however, *one* valid distinction between philosophy and science, even if there are no methodically produced propositions that we can say are solely the property of philosophy rather than of science. This distinction lies in the nature of the person of the critical inquirer and consists in the purpose of the person who troubles himself with methodical inquiry into propositions and statements.

In order to do justice to the historically received phenomenon of philosophy, we must not only compare the propositions of philosophy with those of science but also compare the philosopher with the scientist.

There is a widely accepted distinction between natural scientists and social scientists—I prefer to call the latter cultural scientists. The cultural scientist must participate in his subject, that is, the objects that he investigates. The cultural scientist must be able to empathize; he must be able to replicate in thought the historical behavior of men, whereas the natural scientist can only observe without taking part.

This "taking part" of the cultural scientist, however, can be merely an intellectual game. It may be motivated by the mere pleasure of being able to understand, as well as by various other impulses, which we are today fortunate to have had delineated for us by psychoanalysis and sociology. But I do not want to go into that here. Rather, I want to differentiate from the hermeneutician a certain kind of philosopher, a type that is usually called the existential philosopher. The distinction is based upon a completely different degree of participation in the results of one's thinking. As has been repeatedly demonstrated in the history of philosophy from Socrates to Wittgenstein, the pursuit of philosophy contrasts with the pursuit of a science in that the philosopher, to put it somewhat solemnly, sets his life on the outcome of the game—the game of thinking. The philosopher thinks with the intention of directing his life on the basis of the results of his thinking. When we formulate this as the philosopher does—"What should I do?"—it is likely to be misunderstood within our Western traditions. To us "I should" sounds too much like heteronomous morality.

On the other hand, statements without "should" fall too easily into the alternative category of that which is pleasant or useful. Once a person ceases to find his activities self-evidently intelligible, he must seek his own way whether he wants to or not. The philosopher is the person who has decided to follow only that way that can be justified through thought. Every esoteric theory is thereby precluded, because he can accept only that which can be defended to any being capable of understanding. In practice, however, philosophy can nonetheless become an esoteric theory when the majority of men behave dogmatically, refusing from the outset to have anything to do with the critical question "What should we do?"

At the present time it is precisely among scientists that we frequently encounter dogmatic behavior. We are naturally critical within our own science but do not question the science itself.

If my analysis of methodical thinking is correct, even critical philosophy has at its disposal only methods of thought that are accessible to every generally intelligent individual—methods that can be learned, for example, by anyone successful in a science. That notwithstanding, philosophy cannot be taught, because we cannot force the learner to apply to his own life what he learns. In fact, we cannot even ensure that the learner will notice that what he learns concerns his own life.

It is also possible—I almost want to say unfortunately—merely to empathize understandingly with the philosopher who sets his life on the outcome of his thinking. We can always misunderstand philosophy as simply an intellectual game.

Philosophy understood as the will to set one's life on the outcome of one's thought cannot be taught, and reflection upon methods of thought does not necessarily lead to philosophy. But without methodical thinking neither science nor philosophy is possible.

TWO

Enlightenment and Reason

Happily, it is gradually becoming a habit for people to begin lectures by explaining how they intend to use the words that appear in the title. I cheerfully fall in with this habit—in fact, I even view it as rational.

To be sure, it is important that we not get bogged down in words. It is wasted effort, however, to pay slight attention to words in an attempt to arrive quickly at the issue, for in the end we will have only used words to obscure the issue.

For my topic, then, I must begin with "enlightenment." In the Brockhaus encyclopedia we also find given as the principal signification under the entry "enlightenment": "(1) the Age of Enlightenment, an intellectual movement that arose at the end of the 17th century and remained influential into the 19th. . . . With an optimistic energy and trusting in the rationality of the world and of men, the Enlightenment molded all areas of culture. . . . An essential characteristic was a tendency toward a scientific posture and an active desire for reform." Following these general characteristics, faith in science and desire for reform (particularly in the area of morality and politics—criticism of Christianity was intended above all), the entry goes on to present representative individuals from the movement. In a deliberately rigorous selection I would name Locke and Gibbon in England and Voltaire and the Encyclopedists in France.

I do not consider either Descartes or Hobbes to have been men of the Enlightenment, because neither belongs to the period of the Enlightenment, which only began in earnest under the spectacular success of Newtonian physics. And I exclude both Hume and Rousseau, because Hume was too skeptical of science and Rousseau was too pessimistic in his views on culture—too rapturous over nature.

All in all, I would like to use the word "Enlightenment" as the proper name of an intellectual movement that designates an entire historical period, roughly, the eighteenth century. The phrase, "the century of enlightenment," can then be used as a designation—a designation for the eighteenth century, as long as we are speaking of European history, of course.

We can further specify the movement known as the Enlightenment exemplarily by means of some of its representative figures (e.g., Locke and Voltaire) and then still more precisely by means of counterexamples (like Hobbes and Rousseau). Beyond this, we can further delimit the movement by means of terminological specifications; for example, "only a person who has faith in science, is hostile towards religion, and is disposed to political reform is a member of the Enlightenment."

Terminological specifications of this sort are not arbitrary, for they must fit the exemplary representatives of the movement, although they guide the choice of exemplars at the same time. For example, it is no use for Christian Wolff to have been the foremost German philosopher of the mid-eighteenth century. He is, nonetheless, not a member of the Enlightenment, because he strove for no political reforms as consequences of the new natural science.

In contrast, the specifications of "having faith in science, hostility towards religion, and being disposed to reform" are particularly well cut to the measure of the Encyclopedists. This is the proper name of a group of French philosophers, scientists, and writers who, under the leadership of Diderot, published the *Encyclopédie, ou Dictionnaire raisonné des sciences, des arts, et des metiers* during the period 1751–80. In particular, d'Alembert, Montesquieu, Holbach, and Turgot belong to this group.

Becuase the Brockhaus encyclopedia has only just completed the letter G, it has at least already come to itself. Under the heading of "encyclopedia" we find a historical overview of those works that seek to display the unity of all sciences, from that of Speusippus, Plato's successor as director of the Academy, through the ancient and medieval collections, to the encyclopedia of Diderot, Hegel's *Enzuklopädie der philosophischen Wissenschaften im Grundrisse* (1817), and finally to the Soviet encyclopedia.

The Brockhaus encyclopedia has not yet arrived at the heading "reason," which is more important for my topic. This represents no disadvantage for these remarks, however, because the issue that reason poses for philosophy is "not to prattle on about reason from the outside, but rather really to practice rationality in earnest." So thought Fichte, at any rate.

Yes—Fichte! Exactly. But don't you know that Fichte was a romantic?, some of you will surely now want to ask—namely, all of you who are members of the contemporary intellectual movement that sees itself as the legitimate heir to the Enlightenment. The Vienna Circle (Schlick, Carnap, Popper) exemplifies this movement well. I quote the following from a 1969 lecture by Ernst Topitsch.

> Influential intellectual groups and ideological institutions still continue to cling to world views and forms of self-interpretation that are deeply rooted in prescientific and preindustrial, indeed, archaic traditions.
>
> This applies in no small measure to Germany. As a consequence of Germany's peculiar political, economic, and social fate in the modern era, this "late nation" participated in the Enlightenment and the scientific-industrial revolution not only noticeably later but also in forms other than those of the other West European states. There was lacking a sophisticated nobility as well as a bourgeoisie that, conscious of its economic power, could have demanded and realized political rights. The culture of the (intellectually most advanced) evangelical part of the German linguistic region was formed by unassuming citizenship and above all by the Protestant ministry. Radical enlightenment on the English and French models was scarcely available, and where individuals did carry on the Western intellectual trends it was to seek to build these into a metaphysical-theological interpretation of the world or to vault over the former by means of the latter. This applies in particular to German idealism, which wanted to overcome the "vulgar" and "destructive" Enlightenment as well as the calcified orthodoxy by means of a philosophically ennobled Christianity.
>
> For a long time the speculative-religious intellectual sphere of romantic idealism in its manifold forms and applications remained the ideological wellspring of the German universities and especially of the humanities faculties; even today its effects are still clear enough.

Appealing to Max Weber's doctrine of science free of value judgments, Topitsch believes the matter to be quite simple. After the eighteenth and nineteenth centuries and the scientific-industrial revolution of that period, anyone who still maintains not only that man should engage in natural science or empirical social science (of course, including the writing of history free of value judgments) but also that man can and should, using the capacity known as practical reason, concern himself with the setting of norms (or values) in the least arbitrary manner possible is a romantic, according to Topitsch.

This word "romantic" comes to us from the history of literature. What is meant is the movement that followed upon the Enlightenment and that in the confusion of the time around 1800 longed for a return to the universality of belief found in the Middle Ages. Imagine to yourself the effects of the terror with which the French Revolution ended and the confusion of the Napoleonic wars. "Romantic" means "as in a novel," and what is meant thereby is the world of the medieval romances, which were first written in the colloquial Romance languages.

From this perspective, anyone who desires a world of valid norms (or values) is a romantic. Topitsch further generalizes the reference of "romantic" in that he calls "romantics of the right," those who long for a *return* to such a healthy world (assuming that it once existed). But Topitsch also calls romantic anyone who hopes for a healthy, or merely a healthier, world in the future. The latter are "romantics of the left." All Marxists, all neo-Hegelians, including (and these are the target of Topitsch's polemic) the dialectical philosophers and sociologists of the Frankfurt School (Horkheimer, Adorno, and Habermas), are romantics of the left.

In fact, even Max Born, the famous nuclear physicist, would be a romantic acording to Topitsch. About ten years ago Born said of space travel: "I belong to the generation that still distinguishes between understanding and reason. From that perspective space travel is a triumph of understanding, but a tragic denial of reason."

Born is saying that today the dominion of mere understanding has been established and that he finds this tragic, because as a physicist it is not possible for him to advance and vindicate the claims of reason. Physics is the paradigm of value-free science, the paradigm for the sciences of the understanding. A person who, in contradistinction, supposes, using Kantian terminology, that reason is not merely theoretical (i.e., that it can produce sciences of the understanding) but also and even primarily practical (i.e., that it can set norms and determine the will) thinks "romantically" as far as the Vienna Circle is concerned. If such a person takes his norms from the past, then he is a romantic of the right—relatively speaking, a harmless dreamer. But, if such a person claims to be able to arrive at norms for today that enable him critically to appropriate the past and if he claims that on that basis practical reason should determine our future course, then he is a much more dangerous romantic of the left.

After Hegel, the romantics of the left labeled themselves "dialecticians." I would rather retain Topitsch's term—precisely because in my subsequent remarks I want to defend the idea of practical reason. An evil hour will have come upon this idea when it need fear a polemically formulated terminology.

Even if we want not to prattle on about this matter from the outside but rather to discuss it in earnest, that does not mean we have decided to call ourselves dialecticians or that we have let fall the intended disparagement of the label "romantic of the left."

It still remains for us to show how we can arrive at norms through non-arbitrary discourse and thought. That certain norms are more than mere beliefs, more than mere acts of will—that is the problem at issue in science's so-called freedom from value judgments.

Anyone who wants to look into this problem empirically, that is, by investigating what contemporary scientists believe on the issue, will not make much headway. You will easily ascertain that freedom from value judgments enjoys a great popularity. For example, Georg Picht, the inventor of the German educational catastrophe, writes in *Enlightenment and Revelation:*

> In the second half of the 20th century science and philosophy are undergoing in all areas a return to the empiricism, positivism, and pragmatism of the great thinkers of the 18th century. Locke, Hume, and Condillac are closer to our present way of thinking than are Schelling or Karl Marx. The standard model of contemporary science is not Hegel's encyclopedia but rather the encyclopedia of Diderot, d'Alembert, and their friends.

In my opinion, this simplistic return to the Enlightenment could, with greater justice, be called an "educational catastrophe."

In another place Picht says:

> Up until the death of Hegel philosophy was considered the science of the conditions of the possibility of the sciences as well as of politics and society. As the bearer of the legacy of Plato and Aristotle, philosophy's theme was therefore the facilitation and mediation of science and politics. We have seen, however, that the sciences of the 20th century owe their successes and structure to their emancipation from philosophy. . . . In this sense the emancipation of the individual sciences from philosophy is irrevocable.

I would rather break off the citation here; otherwise, we run the danger of finding ourselves suddenly faced with revelation instead of reason.

If we consider what is generally believed today, it really does seem as if in its triumphal procession value-free science cannot be brought to reason, as if the efforts of thought from Kant to Marx, despite modern physics, to demonstrate the normative power of reason were nothing but wishful thinking, nothing but a romantic fantasy.

The men of the Enlightenment did not encounter this problem, for they lived in innocent ignorance of the distinction between understanding and reason or, what is the same thing, the distinction between theoretical and practical reason. At that time the undeniable success of Galilean-Newtonian physics had led to a faith in science, and with that faith people believed that the state and society could also be known scientifically and reformed using that knowledge. The followers of the Vienna Circle are therefore not men of the Enlightenment, for they lack this naive faith. To be sure, as did the men of the Enlightenment, they want science to be practiced only on the model of the natural sciences, and for that reason they are called "scientistic" by their opponents. Yet they *know*, as the men of the Enlightenment did not, that their science must be value-free and that it can supply no norms for the state or for society. Thus, they continue to fight bravely against pre-Enlightenment obscurantists, of which today there are still enough, but they no longer fight for political changes. If we ask whence they have their "freedom from value judgments," the answer runs, "From Max Weber." But Weber was a neo-Kantian (even if an idiosyncratic one), and his distinction between theoretical and practical reason derives from Kant. In opposition to the scientism of the Enlightenment, Kant wanted to reestablish the primacy of practical reason. We can say *re*-establish, because for Plato, in his opposition to the sophists, the primacy of practice was the reason for setting philosophy in general into motion as an autonomous pedagogical tradition. In Max Weber, however (out of an intellectual honesty of which it is easy to approve), there remains nothing of the teachability of practical reason. So much then for intellectual history—from history alone we obtain nothing concerning our problem. Kant and Hegel, nonetheless, might have been correct in their claim that reason is practical—even if it appears that only a handful of diehards, the Frankfurt School of dialecticians, are still prepared to defend this claim critically.

From the perspective of the Vienna Circle and its scientism the dialecticians are merely vestiges of romanticism. From the perspective of the Frankfurt School scientism is a variant of the Enlightenment, which has already been superseded by the work of Kant and more particularly by that of Hegel. From the perspective of intellectual history it is impossible to reach a decision on the matter. We must look upon the respective assertions of the Vienna Circle and the Frankfurt School as systematic assertions and must determine which is defensible when we put them each to the test of the disciplined scrutiny of our own discourse, as we do in Erlangen.

To that end I present here a sample from Habermas's work on technology and science as ideology.

> The tension which Marx diagnosed between productive forces and social institutions is due to an ironic relation of technology to practice. In the age of thermo-nuclear weapons the explosivity of this tension has grown in a way not previously seen.

The direction of technological progress is still today largely determined by social interests which grow naturally out of the pressure for social reproduction and which as such are not reflected upon and confronted with the clear political self-understanding of social groups. As a consequence, technological ability breaks unprepared into the existing practical forms of life. New potentials for even greater technological capability make manifest the mismatches between the results of rationality stretched to the limit and unreflected upon goals, value systems grown rigid, and failing ideologies.

Today in the most industrially progressive systems there must be undertaken an effective attempt consciously to carry through the mediation of technological progress with practical life in the great industrial societies—a mediation that has previously occurred in the natural course of history. . . . It is a matter of creating a politically efficacious discussion that will rationally bind the social potential of technological knowledge and capability to our practical knowledge and desires.

I hope you are pleased by such amusing formulations as, for example, "ironic relation of technology to practice"—Habermas might just as well have spoken of an "ironic relation" of (value-free) science to politics. In any event, this kind of talk leaves an empty aftertaste, because everything is left merely programmatic. "Today . . . there must be undertaken . . . an . . . attempt consciously to carry through the mediation of technological progress with practical life. . . . It is a matter of creating a . . . discussion that will rationally bind the social potential of technological . . . capability to our practical . . . desires."

Somewhat later Habermas formulated this as "This dialectic of capability and desire occurs unreflectively today, dictated instead by interests . . ."

He *demands* that this dialectic be brought under the control of rational discussion—but according to scientism that is precisely pure wishful thinking. According to the adherents of scientism, rational discussion encompasses only the scientific (that is, it is free of value judgments), the determination of what is *in fact* desired, and the determination of the means to the realization of such actually desired ends.

Herbert Marcuse also exposes himself to the reproach of romanticism when he complains that contemporary man thinks only "one-dimensionally"; that is, he sticks only to the empirical facts and thus no longer entrusts the normative second dimension to his reason.

Horkheimer, the grand master of the Frankfurt dialecticians, speaks more dispassionately. In a 1940 essay, "Die gesellschaftliche Funktion der Philosophie," he wrote:

> Philosophy is the methodical and persistent effort to bring reason into the world; that is what occasions philosophy's precarious and disputed position. Philosophy is disagreeable, obstinate, and, moreover, without immediate use—namely, truly a source of aggravation. Philosophy lacks clear criteria and compelling demonstrations. True, empirical research is tedious, but at least you know what it is about. . . . The fact that theory can evaporate into a hollow and bloodless idealism or into tedious, empty verbal nit-picking does not mean that these forms are philosophy's true forms. . . . In any event, the entire historical dynamic has today placed philosophy at the center of social reality and social reality at the center of philosophy.

So that you are not bedazzled by the glare of rhetorical polish on the last sentence, let me repeat the beginning of the quotation: "Philosophy is the methodical and persistent effort to bring reason into the world." It is the adherents of scientism who claim, however, to know that the only rational possibility is to restrict reason to science that is free of value judgments. The remainder of life's activities are purely a matter of volition, purely a matter of deciding something or other. Herman Luebbe is also a proponent of Max Weber's and Carl Schmitt's decisionism—unfortunately, he left Erlangen too soon. Luebbe formulates this position in the following way.

> Once the politician wanted to be held in exalted respect over the technician, because the latter only knew and planned what the former understood how to accomplish. Now the situation is reversed insofar as the technician understands how to read what the logic of relations prescribes, while the politician advocates positions in the sort of disputes for which there are no earthly courts of rational decision.

This decisionism, which is simply a consequence of the scientistic restriction of reason, is correctly criticized by Habermas when he observes that in it "a disciplined discussion" beyond the borders of scientism's approved way of speaking is "precluded from the outset."

With that I want finally to leave off quoting. The play in the match, Vienna versus Frankfurt, clearly stands at a score of 1:1. It is a fact that the adherents of scientism skeptically restrict reason "from the outset" to theory that is free of value judgments. And it is a fact that the dialecticians merely assert dogmatically that it must be possible for reason to be practical. Both groups claim for themselves the famous Kantian label "critical," but at most only one of the two schools has a chance of being truly critical. I would maintain that it is the Frankfurt School. And I maintain this precisely because only its assertions can be defended—assuming of course that these assertions are so examined that our own discourse is reflected upon critically in the course of the examination.

According to the analysis of intellectual history presented earlier—that Kant's attempt to go beyond the Enlightenment has foundered since Hegel's death—we must try to repeat the Kantian critique. We must, with Kant, demonstrate that our theoretical knowledge does not preclude the possibility of practical reason. We can do this (we could even say "ironically") precisely by demonstrating that theoretical reason itself has a normative foundation. It is only out of and on the basis of the preexisting (even if very inadequately comprehended) significance of everyday human activities that we can specify an initial set of norms for scientific discourse. And it is only on the basis of such norms that it is possible to have disciplined argumentation, for example, the logical inference, calculation, and objective measurement that are the foundation of physics.

Thus philosophy begins with the "methodical and persistent effort" to bring reason into the *sciences*. I cannot give here in detail this first step onto the path that we have followed to arrive here today, but you will have retraced that beginning yourself, if you will take the trouble to read through to the end my *Logische Propaedeutik* (Mannheim, 1967). The physicist who conceives his own activity in this way and who realizes that initially he must *critically* appropriate certain norms of discourse (those norms required for logical and mathematical inference) and certain norms of action (those required for physical measurement) has already refuted scientism in his actions and thought. By his own activity such a physicist has proven that his reason is practical, that his reason possesses normative power. He has thereby also proven the teachability of practical reason, because he can have *critically* appropriated the norms only if he has been able to justify them to other people, including nonphysicists.

I do not want to prophesy what would happen if reason were to be brought into the sciences in this way. Even then a scientist could flee from reason's demand that we critically assimilate the norms that initially give meaning to our activities. These norms could simply be taken up dogmatically. It would be appropriate to ignore such a "scientist," if he nonetheless still put forward claims concerning whether reason could or should be practical.

Now, many of you who are not natural scientists will presumably be inclined to think, "That may very well be—reason should be brought into the natural sciences—but what is the situation with the so-called human sciences, specifically, with the social sciences, for example, the science of economics and political science?" How are we to justify the fundamental specification of goals that initially constitutes these "cultural sciences" (my preference for a collective label) as meaningful activities? Again ironically, the question sounds incredibly difficult, but, on the contrary, once we remove the obstacle of the natural sciences, which has led to the erroneous belief that reason is merely theoretical and not primarily practical, then all that remains is to apply the normative power of reason to the remaining areas of culture. Such areas would be, for example, economics, art, politics, including all social or state institutions like legal services, health services, the construction industry, and so on.

If a philosopher does not want to make a fool of himself by acting as a dilettante in a particular branch of science, he must refrain from making judgments on technical questions. To cite a pertinent example, a philosopher "qua philosopher" (as we say) cannot determine whether it would be desirable at present for West Germany to retain a capitalist economy or to strive for a socialistic economy. He must leave these kinds of judgments to the special branches of science. Of course, this goes only for situations in which the specialists do not practice their specialty in a scientistically truncated form. Max Weber correctly spoke against specialists who taught their opinions and world views dressed up in the scientific clothing of their specialty. It is true that Weber felt it was not possible for reason to become so disciplined that it could methodically (that is, explaining and justifying each step taken) succeed in critically evaluating the norms of an area of culture. But it has always been the scientistic opinion that reason cannot be practical—and philosophy has always countered that reason, however, should be practical.

If we abandon philosophy's counterclaim, cultural matters (state and economic institutions) are then turned over to those interests that, as Habermas puts it, "grow naturally out of the pressure for social reproduction." The emphasis here should be placed on the word "naturally." If we abandon the effort to conceptually comprehend an area of culture, we turn it over to purely natural interests—more precisely, to the barbarism of the raw natural condition. All cultural institutions, whether political, economic, scientific, or aesthetic, are products that man has laboriously erected upon the foundation of his natural needs and his natural abilities. Man has formed himself from a natural being into a cultural being. This process of formation has always been a product of practical reason. With the first appearance of philosophy, Socrates, man had already begun the process of forming his opinions and desires, which are the basis of all activity, out of his purely natural condition and had begun refining this process itself into a cultural product. Today it is an open secret that the efforts of the Platonic dialectic coupled with Aristotelian logic were directed toward constructing from naturally occurring language a tool that could be used to refine this process of forming opinions and desires and taking them out of and beyond their natural condition. That the disciplining of opinions and desires must begin with the disciplining of one's own discourse was precisely the insight to which the Platonic Socrates constantly tried to lead his interlocutors. Although it is true that in the history of philosophy the Platonic dialectic in this sense has not been successfully preserved, this failure proves nothing conclusive against philosophy. Only with modern science, popularized during the Enlightenment, did reason return to itself, if only in the scientistic truncation just described. In German Idealism, that is, in romanticism (particularly the romantics of the left, as Topitsch calls Kant and Hegel), reason returned to itself as practical reason. To be sure, German Idealism passed over language; more precisely, it passed over reflection upon one's own discourse. In my opinion the Frankfurt dialectic still suffers from this lack of reflection. In the Philosophy Department at Erlangen, under the self-imposed constraint of critical discourse, when we take the trouble to reiterate romanticism of the left, the result is that the proposition "Reason should be practical," for example, is seen to be a truth that follows simply from the methodical examination of the words that occur in the proposition—"reason," "practical," and "should." A methodical examination of these words cannot be laid out here. The word "should" alone requires a theory, modal logic, that remains controversial (outside of Erlangen) to this day, although it was begun by Aristotle. Here I can only refer you to my *Normative Logic and Ethics* (Mannheim, 1969).

To be sure, *pure* philosophy, which concerns itself only with the rational specification of the usage of such words as must be presupposed by all of the individual sciences, is a very abstract but, for that reason, narrowly circumscribed piece of pedagogy (an encyclopedia, not in Diderot's sense but rather in Hegel's). Only specialists who have built their science upon such a reflected-upon system of basic concepts and propositions have the possibility (and also the obligation) of arriving at well-founded judgments concerning norms within their specialty and of coming to well-founded value judgments. This last statement is no longer merely a programmatic claim, because anyone, even the adherents of scientism, can appropriate the norms of discourse from which this proposition follows by using his or her own practical reason (and possibly in contradiction of self-understanding). The emancipation of the individual sciences from philosophy, which is irreversible according to Picht, would then be reversible. It now seems to me that the score stands 2:1 in favor of the Frankfurt dialectic against the Vienna scientism.

My topic has been Enlightenment and Reason," and I hope you have expected from me as a philosopher not historical instruction but rather a discussion of one of our real problems, namely, the problem of the scientistic truncation (or one-dimensionality) of reason that arose during the Enlightenment. I further hope that some of you who are currently engaged in science that is free of value judgments have already at least occasionally had the *feeling* that it might be an oversight not to critically concern yourselves with value judgments. And I hope that those of you who have had this feeling have here been given an additional incentive reflectively to rework these feelings into *insights*.

THREE

Political Anthropology

To the great surprise of East and West, there seems to have been realized in the "Islamic Republic" the principle that religion and politics are "inseparable." Strictly speaking, the pope also represents this principle, but in the case of Western churches we have grown accustomed to a separation of word and deed, of theory and practice.

The weaker principle, that ethics and politics are inseparable, also has many proponents outside of religion. But science, the unrestricted intellectual authority of our time, is silent on this particular issue. Science is silent, although in "theories" *of* science the judgment on this is that scientific ethics is as impossible as wooden iron. No one would deny that such an ethics—were it only attainable—would be very desirable. It remains then to remedy by scientific means, at least partially, this lack of direction in the ethicopolitical realm, a lack deplored by nearly everyone.

The question of a scientific ethics requires two things for an answer: We must define what is to be called "scientific," *and* we must define what is to be called "ethics." Such definitions in turn should not be erected arbitrarily but rather should be justified through argumentation.

For a justifiable definition of "science," a bit of theory of science is required. Only after that can we investigate whether there is a problematic that can be named "ethics" and that we can investigate scientifically in the sense of our justified definition of science.

Thus, in the first part of this essay I will argue for a definition of science such that at a minimum mathematics and physics are sciences in the sense of this definition—and such that at the same time room remains for other sciences. In the second part I will argue for a definition of ethics such that its problems can be dealt with scientifically in the sense of the definition given in the first part.

This second task cannot be made intelligible without the first part. Nonetheless, I want to state beforehand the second task's most important result so that the fewest possible false expectations are engendered. To that end I trust that you have some preliminary notion of what one might call an ethically oriented politics, as opposed, say, to so-called realpolitik or power politics. In the second part of the discussion I will argue that a scientific ethics is attainable when we seek a politically oriented ethics, an ethics that makes a scientific politics possible.

Here too I want to avoid a common misunderstanding. By scientific politics I do not mean a science that claims to be competent to make practical political decisions. What I mean is only a theoretical politics that encompasses the principles of politics in a scientific manner (for example, at present these principles are known as basic values). Therefore, in the most favorable cases this theoretical politics is competent to give a more or less long-range orientation to practical politics. I want to avoid giving rise to any expectations of scientific answers to problems of personal life (and in personal life there are many problems that are traditionally referred to as "ethical" or "moral"), because such expectations must be disappointed. It will be seen, however, that a politically oriented ethics penetrates deeply into much that is often considered a private matter.

These anticipatory observations will become clearer in the second part of the discussion. Therefore it is high time that the first part were begun. We must take up the question of a *concept* of science, that is, a justifiable definition of science. If we define a word, then we have thereby defined a concept. For example, in a different language one would use a different word in the definition, but that which remains invariant in the case of such a substitution is the concept. More precisely, that is what is called "concept" in English.

As I indicated earlier, science is to be so defined that at the least mathematics and physics fall under it. To be still more specific, I choose here geometry and mechanics. Why does one, for example, call that which a mechanic does science? Here of course I do not really mean what is meant today by "mechanic," for example, a car mechanic, but rather what was originally meant by the term, a person who studies the mechanics of natural processes, a physicist, a theoretical mechanic. To pull a cork from a bottle is frequently beyond the competence of a theoretical mechanic, but he is nonetheless responsible for the theoretical explanation of this process. The theoretician can calculate the required force but does not need to be able to produce the force.

It is only practical competence that is possessed by the practical person. Nonetheless, every practical mechanic knows that his practice is, as we say, secured by a theory. Mathematics and physics are indispensible as theoretical supports for all technical practices; without their support our technology would collapse. Up to the present no one has taken explicit responsibility for simple technologies—for example, the removal of a cork from a bottle. There still exists today prescientific technology, and this technology was already in existence long before the development of the sciences in advanced cultures. Usually an engineer is brought into a technical activity only when we cannot succeed without him. The engineer drags in a physicist only in emergency cases, and the latter in turn brings in a mathematician only in the direst need.

In our educational practices, however, this sequence that begins with the practice is reversed. You study mathematics first and then in physics apply what you have learned. Ask mathematicians why they engage in their science, and you will only very rarely receive the answer: to provide theoretical aid and support to physics. Physicists in turn are correct in their almost constant denial that they engage in their science in order to give theoretical aid and support to technology. Their goal is not technological utility but pure knowledge of nature.

If we substitute the word "science" for the word "knowledge," to avoid inadvertently falling into a circular analysis, we can see that we have concluded that the only goal of the natural sciences is the science of nature. In the course of history of our culture, mathematics and physics have become ends in themselves.

Because it would be absurd to call every activity that is an end in itself a science (dancing and playing, for example), we have no choice but to oppose this self-understanding of contemporary mathematicians and physicists and to call their respective activities "sciences" only because these activities provide theoretical support to practice, namely, our technological practices.

It has been with a great deal of thought that our technology has developed out of primitive practices; our technology is practice supported by theory. Irrespective of whatever many physicists may understand physics to be, all physical theories, from classical mechanics to quantum mechanics, are useful as theoretical supports of our technology—indeed, unavoidably so. How much of this technology we really want, however, is not a question for the natural sciences but is, rather, a political question.

For many people the use of the adjective "political" is an indication that it has been decided that the question is not a scientific question. But it is precisely that assumption that needs to be thought through once more, if we are to inquire into the possibility of scientific ethics.

There can only be ethicopolitical sciences if (corresponding to technological practice) there can be demonstrated that there are political practices for which theoretical thought is possible—and desirable.

This theoretical thought need not be *demonstrative* like the mathematical and physical thought that is adapted to technology. If the matter is not to be left to just any old thinking, then this theorizing must at least be *argumentative;* every definition and every proposition must be argued for in a way that is capable of creating a consensus. Here a consensus means general agreement, general denial—or general withholding of judgment. Formulated in Roman legal language, consensus meant at that time: *sic, non,* or *non liquet.* Ignoring the non liquet—in English, "not yet clear"—then with this requirement we are demanding of an argumentative science that it answer every relevant question. That, however, not even the demonstrative sciences are able to do. Regarding the question of whether or not there are infinitely many prime-twins, the answer of mathematicians today is: non liquet.

Only a science can afford this seeming luxury of the non liquet—in practice, whether technical or political, things must be decided. The requirement that scientific argumentation produce consensus (in cases of real disagreement, at a minimum a non liquet, that is, a general withholding of judgment) distinguishes this argumentation from any rhetoric the goal of which is to persuade another simply to accept an opinion. If the other person does not agree (that is, if there is real disagreement), then in rhetoric that means only that each person sticks by his own opinion on the matter.

Among scientists, however, real disagreement means that for all participants the problem is still unclear. The problem must be discussed and argued further, and all have to withhold their opinions. Each person may have his personal suspicions or desires. It is part of the discipline of scientific thought, however, to recognize and accept a non liquet in cases of real disagreement. This disciplining of speech into scientific argumentation or demonstration is learned only by learning a science. Therefore, there is no point in my stopping at the point of defining political science as argumentation productive of consensus and able to provide theoretical support to political practice.

For the second part of my discussion I now need to define which practices are to be called "political." In this connection I will also be able to define "ethics." Ethics will be an essential part of politics, and I will restrict myself in this discussion to the principles, the basic concepts and norms, of a political anthropology.

I will begin my definition of "political" practice with two counterexamples.

If (1) a power struggle breaks out within a group of monkeys, that is not political practice, no more than are (2) the "activities" undertaken by a criminal organization. Saint Augustine observed that states are often indistinguishable from criminal organizations. The distinction is supposed to lie in the fact that political practices serve to support and maintain a *normatively ordered life within a community*. The norms that order life within a community need not be "enforceable" in the modern sense, that is, by means of a monopoly of force in the hands of the law—a Germanic moot also provides a normative ordering of communal life, a "political" ordering, as we would say, using the Greek term.

I cannot stop anyone who prefers not to denominate as "political" those practices that support and maintain a normatively ordered communal life. You can then tell me which word you would prefer to use. I do not quibble over words but rather have only tried to show by means of the examples adduced that it is worthwhile to provide a special term for those practices that work toward a normatively ordered life. Using here the phrase "*political* practice," the question concerning the "possibility" of political science, that is, of political theories, now runs: Does political practice admit of support and maintenance by means of argumentative thought that is productive of consensus?

In our society no one is legally bound to wrack his brain over political questions—the maintenance and improvement of normatively ordered communal life in the West today. It would be plausible to suppose that it is political scientists who are at least professionally obligated to do that, but that is not the case. There are no departments or ministries of government that intervene when a political scientist chooses to define himself as an empirical social scientist or simply as a historian.

Now, that is how things actually are with us. The social sciences as well as the technical sciences are widely believed to be "value-free"; it is only out of personal inclination as private individuals that some scientists are politically active. Using educated language we say that such individuals have, in addition to their scientific pursuits, political *interests*. Scholarly, educated language, which confounds inclinations and interests, obscures in this way the fact that there is a rational political interest. The definition of this interest runs: Whoever devotes thought to political practices, that is, takes the trouble of engaging in argumentation productive of consensus, has political interest.

No one can teach physics to a person who does not come with a technical interest (everything else is merely pedagogical technique). In the same way no one can understand political theory unless he brings with him a political interest. Such a person must at least in thought participate in political practice—and try by means of scientific thinking to maintain and to improve that practice.

To link together in this way practice, political practice, interests, and science does not yet answer the question of whether there are, in fact, scientific principles of politics. But out of this definition of political practice as a practice directed toward a normatively ordered communal life it does follow that there can be political principles, if at all, only on the basis of anthropological principles (put traditionally, basic ethical concepts and basic ethical norms). Physics and all technological sciences do not presuppose an ethics. Even monkeys could participate in technological practice—if they could speak (more precisely, if they could do mathematics), they could also be physicists.

Before participating in political practice, however, a person must first be trained to participate in a normative communal order. Norms can only be represented linguistically. In contrast to technological practice, even the most primitive political practice requires a common language for communal life. In more modern terms, a dialogic community is a condition of any political practice—and thus a fortiori a condition of every political theory. Therefore, as a science, ethics begins with an effort to preserve and improve by means of reflective thought the dialogue that serves and promotes communal life, prior to any and all political organizations. I have no hidden investment in the word "ethics"; in no way do I want to "moralize" in the usual sense. "Ethics" serves here only as a term for that bit of science that precedes politics, that is, a *protopolitics*.

I hope I have now said enough in the way of preparation for the methodical construction of the ethicopolitical sciences that is now to begin. We will start with the practices involved in dialogue and then try to improve these practices. Here I am allowing myself to shorten progressively the phrase previously used several times, "preserve *and* improve"—a phrase that still lies on this side of the distinction between conservative (only preserve) and progressive (only improve). Dialogue as we practice it, particularly among theoreticians, seems to me unequivocally in need of improvement.

For dialogue we do not need to have available a language that is a finished cultural product. We can manufacture for ourselves the linguistic means necessary for communal life. Only such a "construction" of the linguistic means at issue (thus the philosophy of science presented here is called "constructive") makes it possible to argue in a way capable of producing consensus; we must also achieve consensus on the linguistic means themselves.

Without considering a specific culture, it is completely meaningless to reflect on which distinctions should be marked out in a scientific conceptualization of political practice. Therefore, there immediately arises the problem of choosing primary words, basic concepts.

People often speak of the choice of a categorial scheme in this connection. One person chooses this categorial scheme, the other person that one—and with this the possibility of coming to consensus is already lost. Instead of a dialogue, each person engages in a monologue directed at himself. In cases of more rigorous and precise use of language, for example, Aristotle's, "categories" are always grammatical distinctions. They are a matter of syntax, the form of language, not of a language's configuration of substantive distinctions. Such categories are transcultural, "ahistorical," as people also say. We do not find such transcultural categories productive of consensus, if we turn to the grammar of cultural languages as they are spoken. Greek science gave birth to this error in that it took Greek as *the* cultural language simpliciter. The scholastics updated this error with Latin, and now modern science (with the exception of the natural sciences) lets itself be guided by a variety of national languages.

Only the modern linguistic critique that began with Frege makes a transnational science of culture-specific politics possible. At any rate, I maintain, the fundamental principals of political theory are transcultural.

This last proposition is demonstrated by carrying through the *empractic* construction of a rational grammar. Buehler's term "empractic" refers to the process of introducing linguistic devices *in the course of and on the basis of* a practical activity. A rational grammar supplies (here I can only sketch this out) at a minimum the distinction between imperative and indicative sentences (interrogatory sentences, i.e., questions, belong to the latter) and the distinction between nouns (e.g., stone, rose, horse, chair) and verbs. The latter are what I call either doing-words (which can function meaningfully in an imperative sentence) or event words (words for natural occurrences not sustained and powered by human action). With regard to doing-words we need to distinguish what I call *action*. All doing that is not action in this sense I want to call *behavior*.

The distinction between action and doing has its basis in the fact that, although doing is common to both men and animals (in contrast to stones and flowers that only grow and are simply moved physically by natural forces like gravity and light), only man talks, and therefore only men question and respond to questions by, for example, responding yes or no—and therefore only men command and respond to commands, namely, by commission or omission. "Action," according to my terminological proposal, is the doing (commission or omission) with which a man responds to an imperative. Instead of responding through action (action is responsible doing), man can also—as do animals—merely react, merely behave.

These three doing-concepts—behave, talk, and action—now lead us into basic political norms—immediately, into basic anthropological norms. In the tradition of Platonic political theory there are four cardinal *virtues*. Christianity then later expanded this quartet to seven virtues by adding the triad faith, love, and hope. The quartet is a product of the writers of the Platonic tradition, because in his anthropology Plato enumerated only three parts to the soul, for which the usual English words are "sensation," "thought," and "will." Today in the West psychologists are the persons with professional responsibility for the last three basic concepts. These words, however, have been taken uncritically out of colloquial speech by psychology, or are avoided completely by behavioral psychologists. The latter option merely produces an example of the natural science of man.

If we take as our basis the doing-concepts delineated previously, it is not difficult to give definitions to these three basic concepts. We have behaviorism to thank for the reduction of talk about sensations, namely, sensations of pleasure and pain, to talk about the *learning* of patterns of behavior. It then becomes analytically true for both animals and men that we seek pleasure and avoid pain when we are simply reacting. It then suffices to talk of a capacity for learning instead of the capacity for sensation traditionally discussed.

As early as the works of Plato we find "thinking" defined as inner speech. Terminologically, I want to let "imagination" stand for so-called representational thinking. In this way all thinking is by definition linguistic thinking. And then all the distinctions that are intelligible for dialogues, for external speech, can be carried over into internal speech.

All practical dialogues begin with *proposals*, which are then deliberated (i.e., considered in the back-and-forth discussion of *argumentation*). In the end, when all goes well, there results a general *resolution*. Noologically (i.e., as regards the activity of thinking), a person begins with *desires* in a conversation with himself. These desires are then *considered* from various perspectives, and in the end, when all goes well, there results an individual *decision*. In cases where this is only a matter of a yes or a no (should I or should I not?) the result is called a *determination*.

When a person comes to a decision or a determination, the person can be said "to will" that which has been decided. The term "will" is in all cases superfluous. It is only a matter of the *capacity to act,* to do what has been decided. If a person has a powerful capacity to act, then such a person is said to be *effective*. This last term is less liable to misunderstanding then the more usual expression "strong-willed." Since the Christian period of antiquity—more precisely, since Clement of Alexandria—people have always spoken of a will that is either good or evil and not simply weak or strong in the commission of good or evil decisions. This Christian will is a relic of the Hebrews' mythical commerce with the God of the Old Testament. In a scientific ethics a distinction between good and evil is meaningful only for decisions, not for the capacity to act. That is a point taken as self-evident by classical Greek ethics. Decisions result from considerations that begin with desires. And we experience desires because, like animals, we react to sensations of pleasure and pain.

Only the capacity for sensation and the capacity for thought are to be judged in terms of good and evil. For good decisions a person then should have a strong capacity for action (effectiveness). Such a person has *integrity*. Traditionally, "is just" was used instead of "has integrity." That is how justice came to be the fourth virtue, although "integrity" did and does refer only to virtuousness in general. Subsequently, reason, understood as the virtue peculiar to the capacity for thought, was differentiated into intelligence and justice, paralleling the distinction between technological and political reason. To preserve the quartet, "justice" was reinterpreted as the virtue of political reason. That is well taken, because in our technological age reason is often misunderstood as mere technological reason, intelligence. The question that remains is: With which yardstick do we want to measure the capacity for sensation and the capacity for reason? How should we handle the case of the man of integrity? I can also ask: How would you rather have it? Perhaps the capacity for sensation should be cultivated for enjoyment (i.e., sensations of pleasure) as much as possible. Pains would then be treated using drugs. The capacity for thought should then be sufficiently technological-rational to bring about the necessary condition of enjoyment. Then the individual's effectiveness, without regard to other individuals, should be strong enough to achieve enjoyment. Yes, with all this we would have most people's ideal: hedonistic success. But in Plato, Aristotle, Thomas Aquinas, and also in Kant we find a much different understanding of the man of integrity. This man does not hearken to his drives, to sensations of pleasure and pain, at the disposal of which he places his capacity for thought only as a means. Rather, he is master of his drives in order first to consider his actions. The relation of the capacity for sensation to the capacity for thought is one of prudent subordination, prudence. Here prudence and effeciveness together create the ethical virtue of courage. Courage is the emotional, that is, nonintellectual, condition for virtuousness. Courage is also possessed by many scoundrels and, when they are technologically proficient, they are at their most dangerous. Contrary to the Judeo-Christian mythos of good and evil will, the distinction between good and evil depends solely on thought.

Where do we find a basis for calling one act of thinking "good" and another "evil" (or, in any case, "not good")?

Logic teaches us that from nothing one can conclude nothing further. We cannot deduce a principle of "good" (or, more precisely, of "rational") thinking out of nothing. It is precisely because of this difficulty that I have defined ethics as protopolitics and set for political thinking the task of serving political practice, that is, maintaining and improving our normatively ordered communal life.

Thus we presuppose participation in political practice and possession of political interests. We have then available the presupposition that our thinking should maintain and preserve our political practices. From that presupposition we can deduce a principle of reason.

In the simplest cases, norms are given by using universal conditional imperatives. Ethicopolitical thinking concerns itself with such norms as for all people are capable of producing consensus. The only alternative to general consensus is the establishment of norms by force, and that is a sociotechnological problem, not an ethicopolitical problem.

In cultures governed by custom and tradition we have tacit recognition of norms, just as children recognize the "natural" authority of their parents, that is, the commandments and prohibitions of their parents. Where traditional authority no longer exists among thinking adults, there remains only *force* as a means of establishing normative order or—the only alternative in the absence of force, that is, in *freedom*—uncoercive agreement arrived at through discussion in which no one is granted authority. The discussion is conducted "without regard to persons." This principle of uncoercive discussion in which all participating *subjects* (the traditional word) become persons precisely by the fact that each does not defend his words simply because they are *his* words—this principle, according to which no one insists upon his mere subjectivity—is called the principle of reason. By definition, a rational subject is a person whose education has produced not subjectivity but precisely mastery over subjectivity. Rational subjects are also called *persons*. In religious language this mastery of subjectivity is called *transcendence*. Reason is then *transsubjectivity*. That transcendence is primarily an ethical principle, not a religious one, is made clear by the connection between religious traditions that embody a principle of transcendence (by which a subject first becomes a person) and those that embody this principle in the form of the transcendent person. Ethics is the rational kernel that resides within the religious shell.

We are now equipped with a basic terminology for the construction of the ethicopolitical sciences. We also have the principle governing all rational argumentation. We will still lack the requisite content, however, as long as we view man only as a Neanderthal or as a child.

To master the problems of communal life we need, besides our technological knowledge, practical training in the political art of using transsubjective discussion to reach unanimity in cases of conflicting norms, that is, the art of reestablishing a communal order. The material of the ethicopolitical sciences is the existing norms of today; the task is to delineate these norms so that suggestions and proposals for change and alteration can be formulated. The task is a critique and reform of existing norms. The principle of transsubjectivity is the criterion to be employed in this critique and reform. Beyond a simple utilitarianism that begins with the fiction of a maximization or minimization of the sum total of pleasure or pain, respectively—as if people never endure pain for the sake of justice—we can suggest, as a direction in which society should change, that people should increasingly become persons constituting a dialogic community in which transsubjective consensus over norms is itself the highest norm. The rational principle of transsubjectivity describes not a substantive condition but rather only a direction. Or, as the Chinese would say, the way is the goal, the Tao. In the modern vocabulary of "basic values" we could say that the pluralism of basic values would be replaced by a universalism with a single "basic value," transsubjectivity (the realization of dialogic reason). Politics would then no longer operate pluralistically as the penultimate while the ultimate (basic values) stood untouchable in the conscience of the apolitical individual. On the contrary, politics would have the ultimate, the basic value of reason, as its direction, and the next step in this direction would be decided in the conscience of the political individual. In his conscience the subject measures himself against his "transsubjectivity." To be sure, the next step is not decided by freeing the individual and society from a pluralism of individually held basic values, but the discussion that will decide the next step acquires thereby a base from which to begin.

It is a very narrow base. As a theorist of science I can only venture to anticipatorily sketch the contribution that would be made by the various sciences, should they accept such a basis.

First, a task of the historical sciences would be to determine and demonstrate that our time satisfies the necessary condition for ethicopolitical reason. The principle of reason (transsubjectivity) has been given justification as the only alternative to force in political associations in which there is no longer a traditional authority preserving the normative ordering of communal life.

This condition was met for the first time in history by the founding of the Greek city-states (from about the seventh century B.C. on). Plato and Aristotle tried to ascertain and demonstrate the principles of ethicopolitical reason and to set these attempts against the sophistical reduction of all political thought to rhetorical technology.

This experiment foundered during the Hellenistic period.

Subsequent to late antiquity and the Middle Ages this condition was satisfied by the second time in history with the end of absolutism. There was at that time no longer a traditional ordering that preserved the normative ordering of communal life within the political associations of Europe, which have become the states of modern Europe.

Our own states have developed out of primitive political associations. In the course of this development these states have developed *primarily* an aesthetico-religious *culture* that has led to a plurality of customary and traditional forms of life. *Secondarily,* these states have produced for the first time in history a unitary technologico-economic *society.* This society requires—and demands—only technological reason. In contrast, the administration of society is for political reason only a means that has the education of the citizen as its purpose; the goal here is the *tertiary* creation of an ethicopolitical *civilization.* A third dimension, the dimension of ethicorational education, is to be added to the two dimensions of the ethicoemotional and the technicorational.

The transsubjectivity that allows men to become persons can be acquired only through a process of education. Education occurs within the forms of our aesthetico-religious culture, in which people *partially* (particularly within small groups) acquire the capability for transsubjective throught and social commerce. No amount of rational thought can substitute for the unitary totality of a cultural form. But rationality demands that no totality be made absolute.

Tolerance that abides others is not enough. It is rather a matter of a social bond that leaves itself open to question until a consensus is found. (In any event, in practice, compromises made in the absence of consensus will always be necessary.) A historically expanded basis for a dialogue concerning a "theoretically supported analysis of the present" that would perhaps result would obviously be still too narrow.

Since late antiquity, rational thought in the West has been reduced to the specifically "philosophical" phenomena of sensation, perception, and thought. The modern reduction of rational thought to technological reason is only a last, absurd consequence of the fact that feeling (joy and sorrow) and thereby trust, love, and hope have been left to the irrational, to religion.

Therefore, we need to examine again the relation of these three anthropological concepts to religion. Such an examination would be an interdisciplinary task for psychology, sociology, and theology.

Again, I can only hazard a sketch here.

In social life we do not act like behaviorist scientists who through the observation of learning processes form judgments about whether others feel pleasure or pain. As more highly developed animals, we have memories in addition to sensations and perceptions. We therefore also have expectations, and by discussing these we are also able to imagine that which is contrary to present fact. Being social animals we have *sympathy* as the natural basis for all life in a community. Shared joy is doubled joy; shared sorrow is sorrow halved.

We would be committing the naturalistic fallacy were we to infer from this a norm, say, the commandment of love. But this natural basis is the necessary condition underlying the fact that in our culture (as well as in all cultures) *trust* occurs at least partially in small groups (families, circles of friends). We expect sympathy and communal effort.

In our culture we have arrived at a subordination, or even opposition, of the "philosophical" virtues of *reason, justice,* and *prudence* and the theological virtues of *trust, love,* and *hope*.

1. *Neither faith nor distrust, only reason* serves as the maxim of secular man.

2. *Trust instead of distrust* serves, on the other hand, for the religious man.

These maxims are presented as creeds; a person *pledges* to live this way (a person "believes"). The opposition expressed in these two maxims shows quite clearly the schizophrenia of our current thinking. Trust is a condition without which no rational thought can develop (not in small groups and even less as the principle of politics).

3. *Without trust, there is no reason.* This is a proposition of political anthropology. No religion can claim this proposition for itself alone. As a pedagogical proposition this proposition reads: Distrust makes a person irrational. Put in the form of a norm (after rational thought is legitimated and justified as a norm) the proposition reads:

4. *Reason instead of distrust.* The religious maxim (*Trust instead of*

distrust) is now seen to be an abbreviation that appears to be "beyond reason" only because reason no longer appears as the middle term uniting 3 and 4.

The secular maxim (1) is by contrast only a position of retreat in which "philosophy" has become divorced from religion. What remains is to understand our Western religious traditions, that is, the history of Christianity, so that the Western churches, which are currently the most important institutions for ethicopolitical education (particularly for moral education), can play their part in the resolution of the miseries of our civilization—a civilization that is threatened with reduction to a technologico-economic society possessing merely a leisure-time culture.

The meanings of the myths of a Creator God and a God of Judgment need to be determined and demonstrated. Since late antiquity the Jewish conception of God (which had already become "imageless") and Greek philosophy have been fused within Christianity. Within this fusion the life of Jesus appears as the exemplary life lived within an unconditional social bond (transsubjectivity).

Good Friday, the death of Jesus, was at the same time the end of the Old Testament God of Judgment, for a god who let his most faithful son die on the cross is not conceivable as a judge. Easter is, therefore, the decisive event in religious history, because the original commune found the strength to live with each other according to the principle of an unconditional social bond without fear of a god of judgment. Since that time Western churches have been institutions for the realization of the principle of transsubjectivity.

On this wider basis, borne in common by Christian and non-Christian, the political sciences (from psychology and sociology to economics and jurisprudence) can begin in a way productive of consensus the common work on the material norms that determine our political *situation*.

PART II

Constructive Logic

FOUR

Protologic: A Contribution to the Foundation of Logic

In the following discussion, "logic" is understood, in the narrow sense of "formal logic," to be the theory of logical inference. The next problem is to define what constitutes a logical inference. A logical inference is a movement from certain propositions (the premises) to a further proposition (the conclusion). We still need to know, however, which of these movements are to be called logical.

To answer this question, let us examine the simplest example of a logical inference, drawn from syllogistic logic. Syllogistic logic is restricted to propositions of the form $X\ R\ Y$, in which X and Y are predicates and R stands for one of four relations a, e, i, o between predicates. Namely,

X a Y	for:	all X are Y
X e Y	for:	all X are not Y
X i Y	for:	some X are Y
X o Y	for:	some X are not Y

It is a fundamental theorem of syllogistic logic—and thereby of logic—that the inference

from X a Y and Y a Z to X a Z

is a "logical" inference. The rule according to which we infer X a Z from X a Y and Y a Z is asserted to be "logically valid." We symbolize this rule, which was called *barbara* in scholastic logic, as

$$x\ a\ y, y\ a\ z \to x\ a\ z$$

and now use the *variables* x, y, z for predicates. It is a characteristic of logical rules that they are not valid for special predicates; rather, they are valid for any arbitrarily chosen predicates—they are *universally valid*.

It has become customary to call logic "formal" because of this universality. Nonetheless, the issue is not whether this term is justified but, rather, with what right we single out the rule barbara together with some others, for example,

$$x \text{ a } y, x \text{ i } z \rightarrow z \text{ i } y,$$

from other rules, for example,

$$z \text{ a } y, y \text{ i } z \rightarrow x \text{ i } z,$$

which are called "logically invalid"—although we can still also "draw inferences" using the latter.

The current opinion is that the basis for this distinction is to be found in semantics, that is, in the theory of the meaning of linguistic expressions. The rule barbara is called logically valid because for all predicates X, Y, Z for which the premises X a Y and Y a Z are true the conclusion X a Z is also true.

In semantics we start with the position that every proposition is either true or false. Rules in which the use of true propositions always leads to true propositions are called *logically valid* rules.

As regards this semantic definition of truth, there remains the question of how we know that a rule has the required property of always leading from true propositions to true propositions. If we pursue this question, we will see that the semantic definition of logical validity assumes the validity of the logical rules themselves. To be sure, the rules are used in a modified form, namely, in metalinguistic form. Instead of barbara we find the assertion that for all predicates X, Y, Z, for which X a Y and Y a Z are true, X a Z is also true. It is very difficult for semantics to refute the objection that the truth of the latter assertion rests upon nothing else but the validity of the rule barbara.

This issue is represented in another example from modern logic. If A and B are propositions, then "A or B"—symbolically: $A \vee B$—is true, precisely if A is true or B is true. For a propositional form $C(x)$ with a variable x for whatever object (individual, predicate, or whatever), the proposition "for all x: $C(x)$"—symbolically: $\bigwedge x \, C(x)$—is true, precisely if $C(x)$ is true for all objects. The semantic justification of the logical validity of, for example, the rule

$$\bigwedge x \, [A \vee C(x)] \rightarrow A \vee \bigwedge x \, C(x),$$

runs as follows. For A and $C(x)$ the premises are satisfied. Thus, for all x, $A \vee C(x)$ is true; that is, for all X, A is true or $C(X)$ is true, *and therefore* A is true or, for all X, $C(X)$ is true; that is, $A \vee \bigwedge x \, C(x)$ is true.

It is clear that the validity of this theorem results only because a corresponding inference in the metalanguage is accepted as valid. If we want to avoid the endless regress that occurs here, the only possibility in my opinion is to give up the semantic definition of logical implication. We must seek a definition that is independent of "truth" and "falsity." Only if we know that certain propositional forms $A_1, \ldots A_n$ logically imply the propositional form A—symbolically: $A_1, \ldots, A_n \vdash A$—can we assert that every variable replacement that satisfies $A_1, \ldots, A_n$ (i.e., that produces true propositions) also satisfies A. We cannot obtain this kind of general assertion, however, without having used logical theorems.

With this negative result we find ourselves pretty clearly in the position of a person who has just successfully sawed off the limb on which he was sitting. Where is there to be found an end to this and a position that offers some prospect of our again ascending the tree of logical knowledge?

In the following I will demonstrate such a position. I want to remark at the outset, however, that we shall not immediately arrive at true logic. We must first scramble through an outlying region—the region of "protologic."

To find our starting point, let us direct our gaze to an—at first insignificant—aspect of logical inference, namely, the fact that every instance of logical inference is a movement from one proposition to another. Every logical inference is an action, an operation. And, in fact, logical inference is a matter of operating with linguistic expressions. To produce from X a Y and Y a Z the new expression X a Z is something that can be learned even by someone who has not the slightest inkling of logic.

In general we shall call any action that is executed according to some rule, according to a rigid schema, a *schematic operation*. To avoid misunderstanding, however, it must be noted at the outset that by no means is every schematic operation to be designated as protological. When a wall is being built, we operate schematically, because the stones are laid one upon the other according to a schema. Nonetheless, that has nothing to do with logic. But we find ourselves headed in the direction of logic as soon as we take schematic operation *as the object* of further inquiry. There is clearly a close relationship between this starting point and Dingler's foundation of the exact sciences.[1] So that we may study examples of schematic operation, let us take, as a substitute for stones and similar things, *figures*, say, o, +, *, &, which are always available in sufficient quantities. These figures are not linguistic symbols but rather are nothing

---

1. Cf. Hugo Dingler, *Das Prinzip der logischen Unabhaengigkeit in der Mathematik zugleich als Einfuehrung in die Axiomatik* (1915); *Philosophie der Arithmetik und Logik* (1931); *Lehrbuch der exakten Fundamentalwissenschaft* (1944).

more than the "calculi," the stones, which were used for counting in earlier times. Our word "calculus" comes to us from these calculi. We will call any operating schema for producing figures a *calculus*.

More specifically, let us stipulate the following. Each calculus has a finite number of *atomic figures* from which all other figures are constructed. In this way we restrict ourselves to linear writing. Further, each calculus has a finite number of rules that we follow in operating schematically. With the latter we restrict ourselves to *conditional imperatives* of the form, "If such and such figures are produced, then make the following figure."

This will suffice for an example. As atomic figures let us take + and o. As a beginning point for the construction of figures let us take +. In addition, we will use two rules:

(1) If a figure A is made, then make Ao,
(2) If a figure A is made, then make $+A+$.

An example of operating according to the schema looks like the following:

```
       +
      +o              (1)
     ++o+             (2)
    +++o++            (2)
    +++o++o           (1)
```

How often and in what order rules 1 and 2 are used is a matter of individual choice.

The *deduction* of figures—which is what these productions according to the rules of a calculus are called—can be quite entertaining, but we do not want to lose any time with it here, because we want to move on to a theory *about* deduction. As a start, then, we shall introduce symbolic abbreviations for the objects of our intellectual examination. A conditional imperative prescribing the deduction of A, if $A_1, \ldots, A_n$ have already been deduced, will be symbolized as $A_1, \ldots, A_n \to A$. Our calculus from above can then be rewritten in abbreviated form as:

(1) $a \to a\text{o}$
(2) $a \to +a+$

The letter a serves here as a variable that is to be replaced by figures constructed from +'s and o's. In general, the description of a calculus consists of:

a list of atomic figures: . . .
a list of variables: . . .
starting points: A_0

.
.
.

rules: $A_1, \ldots, A_n \to A$

Every formula A is constructed using atomic figures and variables. The following is a further example of a calculus:

atomic figures: I, +
variable: a
starting points: I
I+
+I
rules: $a \to +a+$
$+a, a+ \to \text{I}a\text{I}$

Our first calculus will suffice for discussion of all the important points. If a calculus is given, then the simplest question that can be asked concerns the deducibility of a figure. For example, is the figure +++ooo deducible in our calculus? We write ⊢ +++ooo. But the figure ++o+++ is not deducible.

How can we be sure that it is not deducible? Anyone who is really certain on this point should, for example, be willing to accept the risk involved in promising fifty dollars to anyone who can deduce the figure. The source of knowledge concerning the undeducibility of this figure cannot be analyzed without some preparation. The following question is more easily dealt with. Consider the rule $a \to a++$. If we were to add this rule to rules 1 and 2, we would thereby expand our calculus. This is a *genuine* expansion; that is, there are figures that are deducible in accordance with the expansion, for example, ++o+++, that prior to the expansion were not deducible. New, additional figures are now deducible. That is not the case with every expansion. The rule

(3) $a \to +ao+$

produces no additional figures; the expansion by rule 3 is *not genuine*. The reader should ask himself whether he would here promise a sum of money to anyone who could deduce from 1, 2, and 3 a figure that the reader himself could not deduce from 1 and 2 alone. We will call *admis-*

sible any rule that effects no genuine expansion, that does not really make more figures deducible. We will also abbreviate this by ⊢, for example, ⊢ $a \to$ +ao+. The source of this certainty of admissibility can now be analyzed fully. To be sure, this certainty—as is the case with every certainty—is not teachable, but anyone can be led to the acquisition of it. If I am not mistaken, we are here at the source of all that which constitutes logico-mathematical certainty.

To clarify this, let me present the following lesson in protologic. Let us assume the following situation. We have two people, say, Fritz and Hans. Fritz has promised to give Hans fifty dollars, if by following rules 1, 2, and 3 Hans can deduce a figure that Fritz cannot deduce using 1 and 2 alone. How can Fritz prepare himself for this contest so as to avoid losing his fifty dollars? He thinks over the possibilities open to Hans. Hans will have to begin every deduction with +. Then he will make deductions in accordance with rules 1, 2, and 3. If he uses rule 3 at some point, it will look like this,

$$+$$
$$\cdot$$
$$\cdot$$
$$\cdot$$
$$A$$
$$\cdot$$
$$\cdot$$
$$\cdot$$
$$+Ao+ \qquad (3)$$

But, thinks Fritz, then I can also make, without using 3, the step that Hans has made using 3, namely:

$$+$$
$$\cdot$$
$$\cdot$$
$$\cdot$$
$$A$$
$$\cdot$$
$$\cdot$$
$$Ao \qquad (1)$$
$$+Ao+ \qquad (2)$$

Fritz will then practice replacing every deductive step that uses 3 with deductive steps that do not use 3. Fritz learns a procedure for eliminating instances of 3. If he has secure possession of this procedure, he can go into the contest with confidence.

Result: To establish the admissibility of a rule for a calculus it suffices to acquire an *elimination procedure* for the rule. The theory of these elimination procedures—particularly of procedures that are not specially tailored to a specific calculus but rather can be used in general—is the subject matter of what I call protologic. Of course, protological considerations are familiar to every logician and mathematician. It is only a matter of bringing them into the light of consciousness—and of seeing that they are the supporting foundation of our usual logic. Along with Fritz the reader will have had the insight into the admissibility of 3. We could say that with the elimination procedure we have proven the assertion

$$a \rightarrow +ao+.$$

We must discuss two other typical aspects to proofs of assertions of admissibility.

The rule (4) $ao \rightarrow a$ is admissible.

What can Fritz do this time to prepare for a contest? When Hans uses rule 4 for the first time, it will look like this,

```
            +
            .
            .
            .
           Ao              (1)
            .
            .
            .
            A               (4)
```

Ao has already been deduced. Ao can only have been deduced by using rule 1. Specifically, Han's deduction must look like this:

```
            +
            .
            .
            .
            A
            .
            .
            .
           Ao              (1)
            .
```

$$
\begin{array}{c}
\cdot\\
\cdot\\
A \quad\quad (4)
\end{array}
$$

and Fritz needs only to repeat the first section from $+$ to A in order to deduce A without using 4.

It is somewhat more difficult to see the admissibility of (5) $a \to ++a$. At first glance, who would assert $\vdash a \to ++a$? Nonetheless, you can also run the risk here, if along with Fritz you consider the following. It occurs to Fritz that for every one of Hans's deductions using rule 5 it must be possible to deduce $++A$ without 5. Consider a first-time use of rule 5. That means that A is deducible using 1 and 2 alone. Then I will construct parallel to the deduction

$$
\begin{array}{ccc}
+ & & +++\\
\cdot & & \cdot\\
\cdot & \text{the sequence} & \cdot\\
\cdot & & \cdot\\
A & & ++A\\
\cdot & & \\
\cdot & & \\
\cdot & & \\
++A \quad (5) & &
\end{array}
$$

in which I place $++$ in front of each figure.

The beginning of the parallel sequence $+++$ is deducible from $+$ using 2. If in the original deduction there is found

$$
\begin{array}{ccc}
B & \text{then now there} & ++B\\
Bo \quad (1) & \text{is found} & ++Bo
\end{array}
$$

Again, this is a deductive step using 1.

If in the original deduction there occurs

$$
\begin{array}{ccc}
B & \text{there now} & ++B\\
+B+ & \text{occurs} & +++B+
\end{array}
$$

This, again, is a deductive step using 2.

Altogether, Fritz now knows that the sequence is a deduction of $++A$ using only 1 and 2. He has ascertained for himself the admissibility of 5.

With that the lesson in protologic is ended. How a person arrives at certainty concerning assertions of the admissibility of rules is now clear. From a philosophical perspective we should take particular note of the fact that certainty concerning such assertions is only another expression for the certainty that we possess the ability to perform certain actions (here, the elimination of a rule). We have achieved here a complete overview of the relation between action and knowledge—in any event, of the nature and manner of the interpenetration of the two.

The necessary protological investigations can be developed, completed, and formulated as general principles. Here I do not want to go any further with such investigations. Rather, I only want to discuss the relation of protologic to logic.

It is clear that every student of logic will immediately argue as follows regarding the case of the admissibility of 3. According to 1 and 2 we have the rules

$$a \to ao$$
$$ao \to +ao+$$

Because "implication" is transitive, $a \to +ao+$ is also valid.

This would be to interpret the operational symbol $\to$ as logical implication. The transitivity of implication says that, if A implies B and B implies C, then A also implies C.

For protologic we could say that, if a calculus contains the rules $A \to B$ and $B \to C$, then $A \to C$ is admissible. Or, somewhat differently, if we add the rules $A \to B$ and $B \to C$ to a given calculus, then in this expanded calculus the rule $A \to C$ is admissible. We formulate this insight as

Principle 1: $A \to B; B \to C \vdash A \to C$

The certainty of this admissibility does not come from having read logic textbooks in which it says that implication is transitive. We could, however, do the reverse. We could say that we have *proven* Principle 1 by using the appropriate elimination procedure:

$$\text{transforming} \quad \begin{array}{c} \cdot \\ \cdot \\ \cdot \\ A \\ \cdot \\ \cdot \\ \cdot \\ C(A \to C) \end{array} \quad \text{into} \quad \begin{array}{c} \cdot \\ \cdot \\ A \\ \cdot \\ \cdot \\ \cdot \\ B(A \to B) \\ C(B \to C) \end{array}$$

and defining the "logical implication" of rules by saying that a rule R is logically implied by rules $R_1, \ldots, R_n$ when the rule R is admissible for *every* calculus after the addition of $R_1, \ldots, R_n$. For the sake of brevity, assertions of admissibility that are valid for *all* calculi are called logical. In fact, the usual logic can be operatively—that is on the basis of schematic operations—interpreted in this way. With the exception of negation, everything is exactly as it is in the classical theory. For negation we have, in contrast, at first only intuitionistic logic—with which, however, we know that we can justify two-valued logic as at least a fiction.

We should not lose sight of how widely the operative interpretation diverges from the usual interpretation. Indeed, at the beginning there are no propositions that are true or false; we speak of rules for operating schematically. True and false do not exist with respect to such rules. It remains to be shown, for example, how the treatment of disjunction nonetheless leads to the usual principles. For this we begin by using some calculus. Among the atomic figures of this calculus the figure ∨ does not appear. We then expand the calculus with the following two rules:

$$a \to a \vee b$$
$$b \to a \vee b$$

These obviously constitute a genuine expansion. The following is now valid for the expanded calculus:

Principle 2: $a \to c; b \to c \vdash a \vee b \to c$

For a proof we need an elimination of $A \vee B \to C$ similar to the elimination of 4. A deduction that uses $A \vee B \to C$ for the first time looks like this:

$$\vdots$$
$$A \vee B$$
$$\vdots$$
$$C(A \vee B \to C)$$

According to the rules with which ∨ was introduced, there must be prior to $A \vee B$ a deduction of A or of B. If it is A, then $A \to C$ immediately gives C; otherwise, we use $B \to C$.

In the case of an axiomatic presentation of logic—which has no bearing on the problem of a foundation—Principle 2 will be placed at the top as an "axiom" next to the introduction rules for ∨. If we ignore the proof of Principle 2 given here, what distinguishes this principle from an axiom is the fact that we take a risk when we assert the principle. Here also we could imagine a situation in which Fritz promises Hans a sum of money if Hans can produce a calculus in which he can (after expansion with $\vee$) deduce a figure using $A \vee B \to C$ that Fritz cannot deduce using only $A \to C$ and $B \to C$.

In this way insight into logical theorems rests completely on protological considerations. In addition, however, protologic also supplies the foundations of arithmetic. For counting, we need nothing more than the figures that are produced according to the following calculus.

> atomic figures: I
> variables: x, y
> starting point: I
> rule: $x \to x\text{I}$

We can immediately obtain the principle of arithmetic induction via a consideration that is in principle no different from the admissibility proof of 5. In order to formulate it, we examine a calculus in which certain figures $A(X)$ are deducible, if X is one of the foregoing "numbers." For every such calculus,

> Principle 3: $A(\text{I}); A(x) \to A(x\text{I}) \vdash A(y)$

is valid. To prove this we examine

	I		$A(\text{I})$
	.		.
	.		.
	.		.
a deduction of a number y	x	and construct parallel to it	$A(x)$
	$x\text{I}$		$A(x\text{I})$
	.		.
	.		.
	.		.
	y		$A(y)$

To the right is a deduction of $A(y)$ with a starting point at $A(\text{I})$ and using the rule $A(x) \to A(x\text{I})$. The existence of such a deduction is precisely what Principle 3 asserts.

If we think of mathematics in the narrow sense as the region of the logico-mathematical (i.e., minus geometry as a theory of that which is spatially realized), then we can undertake to erect all of mathematics upon the basis of protologic. The mathematician can see that the *operative mathematics* obtained in this way is radically distinct from the usual mathematics. There is no place in it for uncountability in Cantor's sense, because we always operate with figures according to rules. Thus only countability in Cantor's sense remains. Therefore, there now arises the question of whether the uncountability of modern mathematics is anything more than a *façon de parler*.

FIVE

Identity and Abstraction

The following observations are intended to clarify the question of the "existence" of abstract objects by investigating in general how we come to talk intelligibly of abstract objects. "Realist" propositions will be accepted as true: for example, "Numerals are figures that signify numbers"; "Numbers are not figures." On the other hand, no "realist" presuppositions will be made, so that "nominalists" can explain my results as a mere *façon de parler*.

Because the issue is now precisely in which way (façon) we are to intelligibly speak (parler) of abstract objects, "realists" will be able to agree with all this.

1. LOGICAL IDENTITY

In the logical theory of identity one proceeds on the assumption that some language has already been given. Such a language L contains:

constants $\xi, \eta, \ldots$
variables $x, y, \ldots$
formulas $A, B, \ldots$

Variables can occur in the formulas. If $A(x)$ is a formula of L in which x (free), as well as possibly other variables, occurs, then the $A(\xi)$ that results from substitution of ξ for x is always also a formula of L. Formulas that do not contain any (free) variables are called propositions.

Identity is defined by

(1.1) $x = y \Leftrightarrow [A(x) \leftrightarrow A(y)$ for all formulas of $L]$

$\Leftrightarrow$ signifies here definitional equivalence, and $\leftrightarrow$ signifies the logical operator "if and only if."

Using this definition (and formal logic), it is easy to demonstrate the following for identity:

(1.2) $x = x$ (reflexivity)
(1.3) if $x = z$ and $y = z$, then $x = y$ (comparativity)

Every two-valued relation that satisfies these conditions (i.e., reflexivity and comparativity) is called an equivalence relation. Therefore, logical identity is an equivalence relation.

2. IDENTITY OF REFERENCE

The following is the simplest possibility for obtaining languages like those that are presupposed by logic.

For specific objects (e.g., persons), we introduce proper names. Were we to introduce the proper name x for an object ξ, then the proper name x refers to that object ξ. For short, we write: $x\beta_0\xi$.

Propositions about the object ξ are then formulated by uttering ξ's proper name x and then adding a predicate to the proper name. For example, using the proper name Socrates, we get the proposition "Socrates is a philosopher." The referential relation β_0 requires that for each proper name there by exactly one object, that is, one and not two (strictly speaking, of course, this requirement is directed at the introduction of proper names).

The requirement that there be (at least) one object yields

(2.1) $x\beta_0\xi$ for one ξ

We now presuppose that we can distinguish between the objects for which we want to introduce proper names. Distinguishability ($\neq$) is assumed to be an available two-valued predicate. This predicate is not dependent on a specific language; rather, it is specified by the requirement that the following be valid for every proposition of every language:

If $A(\xi)$ and not $A(\eta)$, then $\xi \neq \eta$

We will use the symbol $\equiv$ for indistinguishability; that is,

$\xi \equiv \eta \Leftrightarrow$ not $\xi \neq \eta$

The requirement that a proper name not refer to two objects can then be formulated as:

(2.2) If $x\beta_0\xi$ and $x\beta_0\eta$, then $\xi \equiv \eta$

Identity and Abstraction | 73

For every language S and identity as defined in section 1, distinguishability follows from inequality:

if not $\xi = \eta$, then $\xi \neq \eta$

because from not $\xi = \eta$ it follows that

not $[A(\xi) \leftrightarrow A(\eta)]$ for some A

Thus,

$B(\xi)$ and not $B(\eta)$ for some B

and thereby $\xi \neq \eta$.

Identity then follows from indistinguishability, and the requirement that there be exactly one ξ for $x\beta_0\xi$ also yields:

(2.3) If $x\beta_0\xi$ and $x\beta_0\eta$, then $\xi = \eta$.

If we define identity of reference for proper names in the following way:

(2.4) $x \approx y \Leftrightarrow (x\beta_0\zeta$ and $y\beta_0\zeta$ for some $\zeta)$

then it immediately follows from 2.1–2.2 that this identity is also an equivalence relation.

For a language that at a minimum contains the relation β_0, under the presupposition of $x\beta_0\xi$ and $y\beta_0\eta$, the following holds:

(2.5) $x \approx y \leftrightarrow \xi = \eta$

Proof: (1) If $x \approx y$, that is, $x\beta_0\zeta$ and $y\beta_0\zeta$, then according to 2.3 it follows that $\xi = \zeta$ and $\eta = \zeta$, so that $\xi = \eta$. (2) If $\xi = \eta$, then from $x\beta_0\xi$ according to 1.1 it also follows that $x\beta_0\eta$, so that $x \approx y$.

3. THE IDENTITY OF FIGURES

According to the operative conception of mathematics, in mathematics we do not deal with given abstract objects to which we must refer by means of proper names in order to produce mathematical propositions; rather, we must begin by constructing figures. I will restrict myself here to figures that are composed of strokes, that is, to numerals /, //, ///, . . .

It is not the case that every proposition concerning such numerals belongs to mathematics. Rather, we must first define which figures will be considered "identical." To that end we specify an operation for deciding whether or not two numerals are identical. This operation consists of investigating whether two given numerals could have been constructed

using identical steps. We begin then with the identity of every numeral /
with every other numeral /. A step in the construction of a numeral consists of the addition of a stroke to both numerals; that is, $x/$ and $y/$ are
identical, if x and y are identical. The true identity proposition $x = y$ is
then defined as all formulas that can be constructed by following the rule

$$(3.1) \qquad \Rightarrow / = / \\ x = y \Rightarrow x/ = y/$$

"... $\Rightarrow$ ..." is an abbreviation for "from ... is constructed ..."

On the basis of this prescription for construction the following is valid:

$$(3.2) \qquad / = / \\ \text{if } x = y, \text{ then } x/ = y/$$

We can further (inductively) prove that this identity = is an equivalence
relation. We now place on mathematical propositions concerning numerals the restriction that they must be invariant with respect to identity;
that is, for mathematical propositions we will use only formulas $A(z)$ for
which the following is valid:

(3.3) If $x = y$, then $A(x) \leftrightarrow A(y)$.

Formulas $A(z)$ that posess this invariance may be called, by way of abbreviation, *number formulas*.

In addition to 3.3 the following is also valid:

(3.4) $x = y \leftrightarrow [A(x) \leftrightarrow A(y)$ for all number formulas]

so that in the language of number formulas identity of figure then is in
agreement with logical identity. 3.4 can be proven in a way that leads to a
demonstration that the formula $z = z_0$ is an invariant formula $A(z)$:

If $x = y$ and $z_0 = x$, then $z_0 = y$.

This transitivity is valid because = is an equivalence relation. Then, if
$A(x) \leftrightarrow A(y)$ is satisfied for all number formulas, it follows that

$$z_0 = x \leftrightarrow z_0 = y$$

If we replace z with x, then we get

$$x = x \leftrightarrow x = y$$

and $x = y$ follows due to the reflexivity of =.

In addition to the figures /, //, ///, we now introduce the figures $\tilde{/}, \tilde{//}, \tilde{///}$. If
x is a numeral, then $\tilde{x}$ is the numeral that results from adding $\sim$ above it.

For numerals with $\sim$ written above them we will use $\xi, \eta, \ldots$ as variables.

We define the following:

$$(3.5)\ \tilde{x} = \tilde{y} \Leftrightarrow x = y$$

and also define a relation β by

$$(3.6)\ x\beta\xi \Leftrightarrow \xi = \tilde{x}$$

It immediately follows from these definitions that β satisfies conditions 2.1–2.2 for β_0. Therefore, we read the proposition $x\beta\tilde{x}$ as "numeral 'x' signifies the number x", for example, "the numeral '/' signifies the number /."

If we were to write β instead of β_0 throughout section 1, then signification would thereby become a superordinate concept of reference. (This observation can also be given a "nominalist" formulation by using "predicate" instead of "concept.")

In opposition to the introduction of numbers (i.e., the introduction of $\tilde{/}, \tilde{//}, \ldots$), one can with justification object that this introduction is indeed possible but not necessary: *entia non sunt multiplicanda sine necessitate*. Because there is never a logical necessity for introducing new objects into speech, it must be that *necessitas* here means only so-called practical necessity, expediency.

This expediency first arises when, in addition to the numerals used to this point, $/, //, \ldots$, we introduce still other numerals, for example, 1, 2, 3, ..., that are supposed to signify the same numbers. We then have $\tilde{1} = \tilde{/}, \tilde{2} = \tilde{//}, \ldots$, so that the numbers are no longer figures; the number 1 is neither the figure $\tilde{1}$ nor the figure /, because these figures are not identical.

The situation that arises when nonidentical figures are supposed to signify the same thing is a special case of abstraction (from difference). We must now examine abstraction.

4. ABSTRACTION

Let $x, y, \ldots$ be some sort of objects (not necessarily figures). For these objects let there by a given language L and specifically a two-valued relation $\sim$.

A formula $A(z)$ of L is called invariant with respect to $\sim$, if the following is valid:

$$(4.1)\ \text{If } x \sim y, \text{ then } A(x) \leftrightarrow A(y)$$

We define a relation $\approx$ with

(4.2) $x \approx y$ [$A(x) \leftrightarrow A(y)$ for all invariant formulas]

Obviously, then, the following is valid:

(4.3) If $x \sim y$, then $x \approx y$

As in section 1 for logical identity, it also results that $\approx$ is an equivalence relation.

Accordingly, then, instead of 4.3,

(4.4) $x \sim y \leftrightarrow x \approx y$

is valid only if $\sim$ is also an equivalence relation.

If $\sim$ is presupposed to be an equivalence relation, however, then in fact 4.4 can be proven to correspond to 3.4.

Then it is precisely the case that, if $\sim$ is an equivalence relation, then $\sim$ is the identity that according to 4.2 belongs to $\sim$. This explains why in mathematics we always start with equivalence relations when we have to perform an abstraction (which is therefore called the formation of equivalence classes).

In addition to the variables $x, y, \ldots$ for objects, we now introduce new figures $\tilde{x}, \tilde{y}, \ldots$. We also introduce $A(\tilde{x})$ as formulas only for (with respect to $\sim$) invariant formulas $A(x)$. The language of these formulas is called $\tilde{L}$. We then posit

(4.5) $A(\tilde{x}) \Leftrightarrow A(x)$

If $A(\tilde{x})$ is asserted, then $A(x)$ is thereby asserted, but at the same time this also expresses that this proposition is invariant with respect to $\sim$; that is, it is valid for all y with $y \sim x$.

That this propositional possibility is expedient is testified to by its frequent usage in mathematics and all abstract thinking, for example, in talk of concepts rather than of predicates. The essence of abstraction is precisely this restriction to invariant propositional forms.

We next deal with the figures $\tilde{x}, \tilde{y}, \ldots$ as variables for objects. We call them *variables for abstract objects*. The restriction to invariant formulas determines which propositions can contain them. As in section 1, we can define logical identity by

(4.6) $\tilde{x} = \tilde{y} \Leftrightarrow [A(\tilde{x}) \leftrightarrow A(\tilde{y})$ for all formulas of $\tilde{L}$]

According to definitions 4.2 and 4.5,

(4.7) $\tilde{x} = \tilde{y} \leftrightarrow x \approx y$

is valid. Then, precisely if $\sim$ is an equivalence relation,

$$(4.8) \quad \tilde{x} = \tilde{y} \leftrightarrow x \sim y$$

is also valid. If we use $\xi, \eta, \ldots$ in addition to the variables $\tilde{x}, \tilde{y} \ldots$ in L and if we define

$$(4.9) \quad x\beta\xi \Leftrightarrow \xi = \tilde{x}$$

corresponding to 3.6, then there results

$$(4.10) \quad x = y \leftrightarrow (x\beta\zeta \text{ and } y\beta\zeta \text{ for some } \zeta)$$

corresponding to 2.3. If the objects $x, y, \ldots$ are not figures, it is not customary to say that they "signify" something. Instead, we read the formula $x\beta\tilde{x}$ as "The object x represents the abstract object x."

Some simple examples of this kind of abstraction are:

1. An equivalence relation between pairs of numbers (x, y):
   $(x_1, y_1) \sim (x_2, y_2) \Leftrightarrow x_1 \cdot y_2 = x_2 \cdot y_1$
   The number pair (x, y) represents the rational number x/y.
2. An equivalence relation between formulas (with respect to the variable z):
   $A(z) \sim B(z) \Leftrightarrow [A(x) \leftrightarrow B(x) \text{ for all } x]$
   The formula $A(z)$ represents the set $\varepsilon_z A(z)$.
3. Cardinal equivalence as an equivalence relation between finite sets.
   The set S represents the cardinal number $/S/$.

In all of these cases abstraction creates the possibility of talking about new abstract objects. Thus there is created new talk of objects, and to that extent there is created a new object, namely, the new talk. To say that the abstract objects are newly created is illegitimate if the property of "being newly created" of the old, representing objects is not invariant with respect to the equivalence relation, as in the cases above.

On the other hand, from the presupposition that the old representing objects "exist" we get the result that "existence" is an invariant property with respect to every relation $\sim$. Thus abstract objects "exist."

Looked at in this way, is it not the case that both "nominalists" and "realists" are at the same time both right and wrong?

SIX

Dialogical Foundation of Logical Calculi

Modern logical calculi can be traced back to Frege. Peano and Russell simplified these calculi, and, since Hilbert, they have been the object of a mathematical discipline, metamathematics. Because logic itself dates back to classical antiquity, however, I shall go into the history of logical calculi from eristic logic through syllogistic logic and the axiomatization of geometry. This will serve as an introduction to my theme, which is the dialogic justification of logical calculi.

In the body of this paper I will briefly describe the three types of logical calculi that have been suggested by Hilbert, Gentzen, and Beth, and then I shall take up in detail the dialogic introduction of logical operators and the justification of general dialogic rules.

The concluding section will comprise a discussion of the so-called semantic completeness of logical calculi, including modal calculi.

I begin my historical introduction with the fifth century B.C. in Athens. The discussion-loving citizens of this city in fact paid money to individuals who offered to refute anyone who was prepared to answer questions with either yes or no. These individuals were professional practitioners of the eristic art.

The practitioner of this art began by putting an opening question: A or not-A.

The responder could choose. If he chose, say, not-A, then he had to answer further questions of this kind until the practitioner of the eristic art had together a sequence of answers, say, $B_1, \ldots B_m$—and then, before the eyes of the public, he would establish that whoever had agreed to B_m *must* also agree to A. The original choice, not-A, was refuted. We conceive of this performance by the practitioner of the eristic art as his having found agreement with the premises B_m, from which A followed *logically*.

It was Aristotle's achievement, however, to observe that in the conclusion of this eristic dialogue—namely, in the apparently self-evident necessity of the movement from B_m to A—there is embedded an entire system of rules, the syllogisms.

By modern reckoning there are twenty-one syllogistic rules that lead from two premises of the sort S a P, S ā P, S e P, S i P, S o P, S õ P to a conclusion of this sort.

The logic of junctors that was developed in the stoa comes to us only in fragments, but it is certain that neither the Aristotelian syllogistic logic nor the Stoic logic of junctors was sufficiently developed to give a serious, purely logical demonstration of geometrical theorems starting from the Euclidean axioms and definitions.

It is only since Hilbert and Pasch that we have had a serious axiomatization of geometry. They began with the position that an area of reality (in this case spatial figures) becomes the object of a mathematical theory when certain fundamental concepts and theorems concerning this area are presented in a nonmathematical manner (e.g., so-called intuition or experience), but all further concepts are explicitly defined and all further theorems are logically deduced. We call this the axiomatic method. A critique of this method or, rather, of its claim to be *the* method for mathematically conceptualizing an area of reality is not the thesis of this discussion. Thus I will only record here my deep doubt concerning the reasonableness of that claim.

The axiomatic method enjoys extensive popularity despite such doubt on the part of constructivists. It is important for logic that the axiomatic method presupposes logic to be a system of deductive rules; indeed, theorems are supposed to be deduced from axioms according to *logical* rules.

According to an observation by T. Kuhn (which is completely apposite in this connection), scientists are inclined to carry over into other areas methods that are recognized as successful in one area. Thus it is that the axiomatic method of geometry has also become a paradigm—and thereby has even been carried over into logic itself. It was this that led to the formalization of logic, to logical calculi.

Obviously, it would be senseless to want to deduce *logically* the theorems of logic from certain axioms. The paradigm was therefore so modified that certain theorems of logic were posited as axioms at the beginning, and the remaining theorems must then be formally deduced from the former. Although the expression "formal" comes from metaphysics, here it means only that the deductive rules may not appeal to the signification of the signs used; a deduction in this sense involves only operations with written figures that have no signification. What, then, are the "theorems" of logic? They are the "logical rules" as they are used in axiomatic theories, for example, geometry. It is logic's job to bring these rules into a surveyable system—and to provide a justification of these rules. Producing a surveyable system is done by constructing logical calculi. Mathematical logic restricts itself to such construction work; the justification problem is left to so-called philosophical logic. As I indicated, I will be going into philosophical logic in greater detail. Next I must briefly present the three most important types of logical calculi. I have named these types after Hilbert, Gentzen, and Beth, respectively.

Logical rules regulate the movement from a given set of propositions to a further proposition, in eristic logic the movement from the premises B_1, ... B_m to A. If we bind the premises together using conjunctions, $B_1 \wedge B_2 \wedge \ldots \wedge B_m$, then we have a movement from a single proposition B to a single proposition A. If this movement occurs according to a logical rule, then we say that B implies A. I write this $B < A$. A proposition A that can also be logically inferred without premises is said to be logically true. If we let $\vee$ stand for an empty conjunction, then A is logically true precisely when $\vee < A$.

If we have available as a logical operator a "conditional" $\rightarrow$, then $B < A$ is valid precisely when $\vee < B \rightarrow A$. The Hilbert *type*[1] of logical calculus has its heart set on formally deducing as theorems of a calculus logically true propositions.

A set of logically true propositions are posited as "axioms" at the beginning, and from these one performs deductions using only the rule of modus ponens (B ,, $B \rightarrow A \Rightarrow A$). For example, for the logic of conditionals Frege's axiom system runs:

$$A \rightarrow B \rightarrow A$$
$$C \rightarrow B \rightarrow C \rightarrow B \rightarrow A \rightarrow C \rightarrow A$$

It is precisely in its subtlety that it is most clear that the Hilbert type is not a very happy solution of the problem. Under the slogan of a "natural" logic, Gentzen[2] proposed to deduce not the logically true propositions but rather the logical implications

---

1. David Hilbert, "Die logischen Grundlagen der Mathematik," *Mathematische Annalen* 88 (1923): 151–65; reprinted in *Gesammelte Abhandlungen* (Berlin, 1935), 3:178–91; see also, Hilbert, "Axiomatisches Denken," *Mathematische Annalen* 78 (1918): 405–15; reprinted in *Gesammelte Abhandlungen* 3:146–56.

2. Gerhard Gentzen, "Untersuchungen ueber das logische Schliessen," *Mathematische Zeitschrift* 39 (1934/35): 176–210, 405–31; reprinted in K. Berka and L. Kreiser, *Logik-Texte* (Berlin, 1973), pp. 192–258.

$$B < A$$

Gentzen wrote these as *sequences* in which at times he represented B as a conjunction,

$$B_1 \wedge \ldots \wedge B_m < A$$

and at other times he also represented A as a disjunction,

$$B_1 \wedge \ldots \wedge B_m < A_1 \vee \ldots \vee A_n$$

We find specific sequences given as axioms at the beginning. It suffices, for example, to give $B \wedge c < c \vee A$ (with c being a primary proposition). The rest consists of formal deduction rules that govern the movement from sequence to sequence. For example, for implication we have

$$\frac{C \wedge B < A}{C \quad < B \rightarrow A}$$

$$\frac{C < B \quad C \wedge A < D}{C \wedge B \rightarrow A < D}$$

Frankly, it is in precisely this context that the expression "natural" strikes me as rather artificial. However, the point of the Gentzen type of logical calculus lies in the fact that in its rules the premises always consist of partial formulas of the formulas that appear in the conclusion.

This property, the property of being a partial formula, does not hold under the cut rule

$$\frac{C < B \quad B < A}{C < A}$$

Gentzen's *fundamental principle* states, however, that this cut rule is admissible; that is, it may be used *as if* it were a rule of the calculus, but it is not contained in the definition of a sequence calculus. Since Beth's work,[3] the custom has been to read these sequence calculi upside down. Instead of looking for a deduction that begins at the top with axioms and, through deductive steps, finally reaches at the bottom a specified sequence, we start at the top with the sequence and then "develop" it downward through all branchings until we finally reach the axioms. According to Beth, these developments are more appropriately written out as tableaux.

3. Evert W. Beth, *Formal Methods: An Introduction to Symbolic Logic and the Study of Effective Operations in Arithmetic and Logic* (Amsterdam, 1958; Dordrecht, 1962).

Instead of starting with the sequence $B_1 \wedge \ldots \wedge B_m < A$, we begin a two-part tableau with the *position*

$$\begin{array}{c|c} B_1 & \\ \cdot & \\ \cdot & \\ \cdot & \\ B_m & A \end{array}$$

Each step in the development then develops only *one* of these formulas, and the remaining formulas need not be repeated. For example, if the steps in the development of a conditional are now

$$\begin{array}{c|c} \Gamma & B \to A \\ B & A \end{array}$$

where Γ (instead of C) stands for a column of formulas, and with a branching

$$\begin{array}{c|c} \Gamma\,(B \to A) & D \\ \mid\ \ A & B \mid \end{array}$$

$\Gamma(B \to A)$ here stands for a column of formulas in which $B \to A$ occurs.

Instead of the deducibility of the sequence, we now have the *closedness of the position*; The point of departure is then precisely a logical implication, if there is *one* development that is closed in *all* branchings. A branch is said to be closed, if a prime formula c appears both to the left and to the right of $\|$. This corresponds to the "axiom"

$$B \wedge c < c \vee A$$

These development calculi supply us with suitable decision procedures for propositional logic and with a suitable approach to so-called semantic completeness for quantification logic.

We can view these types of logical calculi as merely formal variants—which is the usual view in mathematical logic. However, Gentzen's expression "natural" goes beyond that. In the following I will show that a dialogic justification in fact singles out development calculi—singles them out as "rational," which I prefer to "natural." To show this, I must first demonstrate how we can justify as rational the fact that we generally use certain operators (in fact, logical operators) in the formulation of propositions, that is, sentences that can be asserted as true.

If we persist in using such words as "rational" and "justify," we will immediately encounter a thousand intellectual arguments to the effect that without logic we cannot "justify" anything, that without logic we cannot single out anything as "rational."

Nonetheless, all these arguments simply produce a verbal fog into which our problem vanishes—namely, the problem of constructing our linguistic tools in a way that can be systematically examined, tools with which we, as scholars, can then argue untiringly and at our pleasure (no matter to what end, no matter against what).

There is only one *nonverbal basis* that can serve as a point of departure for such a methodical construction: our practical nonverbal activity.

In the context of a specific practical activity any normal person can learn how to use sentences of, say, the form

$$N \pi P \text{ or } N \varepsilon Q$$

with a proper name N, an action word P, or a thing word Q, and the copulae π and ε (in English: "do" and "is," respectively).

We learn this kind of sentence and the words that appear in them exemplarily. In this way we have a speech practice that is "justified" within the context of practical activity. This is what Buehler called *empractic* justification. Only by participating in an activity do we acquire the speech "appropriate" to that activity. We learn by practice what it is to assert propositions or to contest the affirmation or denial of propositions (e.g., by nodding or shaking one's head). We introduce a negator, ⊣, where ⊣a is used to express that we are contesting the proposition a. So long as we learn the usage of elementary propositions only empractically, that usage will obviously vary greatly from person to person. It then seems desirable to normalize usage. For example, someone asserts that a thing N is a "fly"; another person "calls" it a "beetle."

Many of the general propositions that are used as examples in syllogistic logic can be justified as normalizations of this kind of varying linguistic usage. "Flies are not beetles" has the form S e P and is to be understood as a *prohibition* against putting forward a proposition $N \varepsilon P$, if the proposition $N \varepsilon S$ has already been affirmed.

Correspondingly, certain general propositions of the form S a P are justifiable as a *prohibition* against contesting a sentence $N \varepsilon P$, if the proposition $N \varepsilon S$ has already been asserted.

To put it differently, the rule of transition from $N \varepsilon S$ to $N \varepsilon P$, $N \varepsilon S \Rightarrow N \varepsilon P$ leads to an incontestable conclusion, *if* the premises are affirmed. Rules of transition in which we must affirm the conclusion if we have affirmed the premises are not logical rules. They are *prelogical;* they provide a set of practical linguistic activities, a set of linguistic practices, which, under rather complicated circumstances, justify the introduction of operators invented expressly for these linguistic practices, that is, logical operators.

Elementary procedures for calculation provide a second region in which we find such rules of transition. For example, in the simplest numerical notation, I, II, III, . . . , addition proceeds according to the following rules:

$$\frac{m + |}{m\,|}$$

$$\frac{m + n}{p} \Rightarrow \frac{m + n\,|}{p\,|}$$

Even at this stage we will require some logic if we want to convince ourselves that, for every result

$$\vdash \frac{m + n}{p} \quad (\text{i.e.,} \ \frac{m + n}{p} \ \text{is deducible})$$

reached by using these rules, it is also always the case that

$$\vdash \frac{n + m}{p}$$

In trying to formulate the *admissibility* of the rule

$$\frac{m + n}{p} \Rightarrow \frac{n + m}{p}$$

we can see the need for a conditional, →, as our first logical operator.

The assertion

$$\vdash \frac{m + n}{p} \to \vdash \frac{n + m}{p}$$

has an employment in which, if a challenger (opponent) can prove

$$\vdash \frac{m + n}{p}$$

then the asserter (proponent) must prove

$$\vdash \frac{n + m}{p}$$

By using the universal quantifier $\wedge$, we can still make it clear that the choice of the numbers m, n, and p is left up to the opponent. As linguistic practices, these rules of employment for → and ∧, which are first formulated in English in deontic propositions, are obviously nonetheless independent of whether one understands English—and of whether one understands deontic modalities in particular. As with every meaningful activity, we can learn these linguistic practices through practice alone, that is, empractically.

At this point one usually refers to the fact that children can learn the rules of a game, say, Chinese checkers, without having previously learned to use deontic propositions. Although that is correct, it is also misleading insofar as it can tempt us to understand logical dialogue rules as mere rules in a game. However, logical dialogues are a rational linguistic practice; only in school, including college, are they turned into "dialogue games." When we arrive at logically composite propositions using *conditionals* (beyond the mere negation of elementary propositions), we can then easily provide justifications for conjunction and disjunction. If we have several conditionals containing the same antecedent,

$$B \to A_1$$
$$\cdot$$
$$\cdot$$
$$\cdot$$
$$B \to A_n$$

then there is an obvious economy in being able to write

$$B \to A_1 \wedge \ldots \wedge A_n$$

instead. This can be done by introducing *conjunction* in the following way. At the opponent's pleasure, the proponent of the conjunction $A_1 \wedge A_2$ must assert A_1 and A_2 as well. Similarly, several implications with the same consequent $B_1 \to A_1 \ldots, B_m \to A$ can be combined into $B_1 \vee \ldots \vee B_m \to A$ by introducing the *disjunctor* ∨. The rule of employment for ∨ looks like this

Opponent	Proponent
	$B_1 \vee B_2$
?	$B_i \qquad (i = 1, 2)$

where the choice of i is made by the proponent. The dialogic definitions for ∧ and →, respectively, are written in this notation as

	$A_1 \wedge A_2$
$?_1 \mid ?_2$	$A_1 \mid A_2$

with a branching showing the choices allowed the opponent and

$$
B? \quad \Big|\Big| \quad \begin{array}{c} B \to A \\ A \end{array}
$$

Negation of specific propositions is given dialogic definition by

$$
A? \quad \Big|\Big| \quad \neg A
$$

Finally—using variables—*quantifiers* are given dialogic definition by

$$
1?\mid 2?\mid\ldots \quad \Big|\Big| \quad \begin{array}{c} \bigwedge_x A(x) \\ A(1)\mid A(2)\mid\ldots \end{array}
$$

with potentially infinite branching, and by

$$
? \quad \Big|\Big| \quad \begin{array}{c} \bigvee_x B(n) \\ B(n) \end{array} \quad (n = 1, 2, \ldots)
$$

with the choice of n left up to the proponent.

The *dialogic definition* of three junctors, two quantifiers, and negation is achieved using *one* rule for each. These six rules can be justified on the basis of any linguistic practices that use rules of transition. These linguistic practices can in turn be justified on the basis of practical non-linguistic activities.

On the other hand, a sequence calculus requires precisely two calculus rules for each logical operator. This doubling occurs because in the case of multiple composite propositions the process does not end in a single column. Using implication and negation, the opponent also asserts composite propositions. The opponent then takes over the role of the proponent. In this way, we hve a second dialogic employment for each operator, if we formulate how the *proponent* is to defend his assertions.

I assume the case of negation to be simpler, because when the opponent takes over the role of assertion on the basis of negation, the proponent does not simultaneously take on a new assertion, as happens in the case of a conditional. In the absence of conditionals, negation creates a pure exchange of roles. The dialogue from that point on looks like a negated thesis of the proponent

$$A_1 \wedge A_2 \quad || \quad ?_i$$
$$A_i \quad\quad\quad\quad (i = 1, 2)$$

and

$$B_1 \vee B_2 \quad || \quad ?$$
$$B_1 | B_2 \quad\quad |$$

with branching, respectively.

Similarly, considering the universal and existential *quantifiers* as "large" conjunctions and disjunctions, respectively, we obtain

$$\bigwedge_x A(x) \quad || \quad ?n$$
$$A(n) \quad\quad\quad (n = 1, 2, \ldots)$$

and

$$\bigvee_x B(x) \quad\quad || \quad ?$$
$$B(1) | B(2) | \ldots \quad\quad | \quad | \ldots$$

with potentially infinite branching, respectively.

For negation occurring to the left of ||, we have

$$\neg A \quad || \quad A?$$

Dialogues for any given composite proposition—assuming that no conditionals occur in it—are completely regulated by these development steps.

Thanks to K. Lorenz,[4] these dialogues are called "strict" dialogues. A thesis is said to be strictly true if it can be defended in a strict dialogue in such a way that it finally leads in all branches to true primary propositions to the right of || or to false primary propositions to the left of ||.

As regards the introduction of logical operators, we are not yet speaking of *formal* logic. We must next consider *material* dialogues dealing with logically composite propositions. Primary propositions result from prelogical linguistic practices.

---

4. Kuno Lorenz, "Arithmetik und Logik als Spiel" (Ph.D. diss., Kiel, 1961); partial reprint in P. Lorenzen and K. Lorenz, *Dialogische Logik* (Darmstadt, 1978).

Strict dialogues reformulate in a pragmatic context that which appears as *semantics* in the usual textbooks—namely (e.g., for conjunctions), that

$A_1 \wedge A_2$ is true precisely when A_1 *and* A_2 are true.
$A_1 \wedge A_2$ is false precisely when A_1 *or* A_2 is false.

Although this semantics presupposes that we already have available a metalinguistic "and" and a metalinguistic "or," dialogic definitions avoid the circular anticipation of a metalanguage. In contrast, the dialogic practice is a new practice acquired through drill, a practice that is in addition to the linguistic practices that produce elementary propositions. As I hope I have convincingly demonstrated, in general the first step that meaningfully expands upon an elementary linguistic practice is the addition of rules of transition. In a second step, these rules justify the introduction of the conditional as the first logical operator after the negator. In contrast, the small and large conjunctions and disjunctions are secondary.

Therefore, we need to expand strict dialogues to include the *conditional*. If, as is usually done in classical logic, we define $B \to A$ by means of $\neg B \vee A$, then the problem disappears, to be sure, but it has not been resolved. In classical semantics this problem does not occur, because we always add *tertium non datur* to it as a dogma. The assertion that for every material proposition either A or $\neg A$ is strictly true is obviously nothing other than a dogma. In the context of junctor logic, tertium non datur is indeed provable, if we restrict ourselves (as is perfectly rational) to decidable primary propositions. As Brouwer first said in 1907,[5] however, this proof breaks down when it comes to quantification over infinite variable domains.

This critique of tertium non datur already indictes that the addition of the conditional to strict dialogues will lead us into the controversy between effective (intuitionistic) and classical logic.

As strict dialogues are defined here, such dialogues have no positions $B \parallel A$ in which formulas stand left *and* right of $\parallel$. Therefore, the dialogic definition of the conditional

$$\begin{array}{c|c} & B \to A \\ B? & A \end{array}$$

---

5. Luitzen Egbertus Jan Brouwer, "Over de grondslagen der wiskunde" (master's thesis, Amsterdam, 1907); cf. Brouwer, "De onbetrouwbaarheid der logische principes," *Tijdschrift voor wijsbegerte* 2 (1908): 152–58.

is applicable only after it has been decided how such dialogues are to proceed.

The recently published book *Dialogische Logik* documents how K. Lorenz and I have labored for about fifteen years to find a "rational" general dialogic *rule*. In retrospect, I can say that we pursued many leads that turned out to be dead ends, but finally we found a road that is open in our opinion.[6]

The basic idea is now very easy to formulate. Because strict dialogues are so strict that in them no conditionals are used, a general dialogue rule must *liberalize* the hitherto strict (too strict) course of the dialogue.

In a strict dialogue, every move always answers the immediately preceding move of the dialogue partner, be the answer an attack (with a question mark) or a defense against an attack. This binding to the immediately preceding move needs to be liberalized.

This formulation is obviously very unspecific. To narrow the choice among possible liberalizations, I propose the following additional requirements, which seem "rational" to me.

1. The liberalization should be *conservative;* that is, all propositions that are strictly true should remain defensible.
2. The liberalization should be *consistent;* that is, for no proposition should both A and $\neg A$ *be defensible at the same time.*

It should be noted here that the strict dialogue rule is trivially consistent in this sense: If A is defensible, then $\neg A$ is not defensible, because the opponent need only undertake the defense of A. This consistency utilizes the symmetry (for both the proponent and the opponent) of the strict dialogue. This symmetry rests upon the fact that the proponent's theses (which contain no conditionals) always lead only to further theses (for the proponent or opponent). In eristic logic, however, the beginning point of a dialogue was that the opponent agreed to produce specific propositions as *hypotheses;* the rules for logical operators then served in the deduction of a thesis (in the case of eristic logic, the deduction of the negation of the answerer's previously chosen thesis). The strict dialogue rule gives no regulation for the development of dialogues that begin with hypotheses and a thesis

---

6. Paul Lorenzen and Kuno Lorenz, *Dialogische Logik* (Darmstadt, 1978); cf. also Paul Lorenzen and Oswald Schwemmer, *Konstruktive Logik, Ethik, und Wissenschaftstheorie* (Mannheim, 1975), pp. 68, 78.

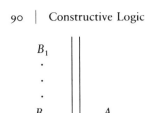

The opponent has only to be able to answer for a thesis. In contrast, the dialogue rule must be so liberalized that the proponent is able to attack previous moves made by the opponent.

Clearly, the opponent will hold to the strict rule that he may answer (in attack or defense) only the immediately preceding move by the proponent.

Not only is this regulation quite obvious; in addition, it has the virtue of satisfying the requirement of conservativeness. If we liberalize the dialogue rule only for the proponent (i.e., we only allow to him *more* moves than are allowed in the strict dialogue), then a fortiori under the liberalized regulation he can also defend every thesis that he could strictly defend.

The simplest liberalization that allows the proponent to counterattack is that which allows him to counterattack at his pleasure. The general dialogue rule that allows the proponent in any move to attack any of the opponent's previous propositions is called the *effective* dialogue rule. As it turns out, effective dialogues lead precisely to intuitionistic logic (I am avoiding the expression "intuitionistic" only because in my justification of this dialogue rule I have not appealed to any of the reasons that Brouwer and Heyting advanced for intuitionistic logic).

Conditionals are handled as follows in effective dialogues. If a conditional appears as the thesis, then we have the following dialogic definition of $\to$:

$$\begin{array}{c|c} \Gamma & B \to A \\ B? & A \end{array}$$

If a conditional appears as a hypothesis

$$\begin{array}{c|c} \Gamma\,(B \to A) & D \end{array}$$

then following an attack by the proponent (B?) the opponent may choose whether to assert A or attack B. This leads in turn to branching

$$\Gamma\,(B \to A) \quad \| \quad D$$
$$|\ A \qquad\qquad B?\,|$$

That is, it leads precisely to the Beth form of the rule for $\to$ used in the "natural" sequence calculus.

The decisive step in the dialogic justification is, however, still lacking. Because we have dispensed with the symmetry between the proponent and the opponent, consistency is no longer trivial. We must now prove that for an effectively defensible thesis A it is never the case that $\neg A$ is also effectively defensible. This result no longer follows from the fact that for the hypothesis A the opponent can simply take over the proponent's defensive strategy for the thesis A, for in effective dialogues the opponent no longer has the same freedom of repeating his attack that the proponent has.

In contrast to intuitionism, the consistency of the effective dialogue rule must be proven in dialogic justification. We do this using a *variant* of Gentzen's *main theorem:* The rule

$$\Gamma \parallel B\,,\,\genfrac{}{}{0pt}{}{\Gamma}{B} \quad \| \quad {}_A \Rightarrow \Gamma \parallel A$$

is admissible for material effective dialogues. For the empty system Γ and for $\wedge$ (false), this cut rule contains, instead of A, the special case

$$\parallel B, B \parallel \wedge \Rightarrow \parallel \wedge$$

That is, if B and $\neg B$ were effectively defensible, then so would be every elementary proposition. Thus B and $\neg B$ are never both effectively true at the same time.

I do not want to go into the proof of this cut rule here.[7] It seems to me more important to try to end the battle between the intuitionists and the classicists by *also* making the case for a classical dialogue rule, now that I have given the above dialogic justification of effective dialogues.

Once we admit conditionals, hypotheses can appear on the opponent's sides, while at the same time a new thesis also appears. We could say that permission to attack these hypotheses before undertaking the defense of the thesis is part of the meaning of a conditional. It does not strictly follow that the effective dialogue rule allows additional repetitions of these attacks as desired. This is justified, however, by the fact that consistency can also be demonstrated for the simplest liberalization of permission to

---

7. Lorenzen and Schwemmer, *Konstruktive Logik*, pp. 84 ff.

attack. The fact that without this dialogic justification effective logic has already arisen as "intuitionistic" logic also seems to me to be an indication of the rationality of this dialogue rule.

Nevertheless, there is no objection to asking why in the liberalization of the proponent we restrict ourselves to allowing repeat attacks. We could further allow the proponent to repeat any move—thus, both attacks and defenses—as long as any further liberalization is also consistent.

In fact, a general dialogue rule that allows the proponent at any time not only to attack the opponent's earlier propositions but also to defend himself against earlier attacks is consistent. This dialogue rule is called the *classical dialogue rule,* because, at it turns out, it yields precisely classical logic.

The proof of consistency is again obtained using a cut principle.

In addition to strict truth (defensibility), we have justified two further concepts of truth (defensibility) for logically composite propositions: effective truth (defensibility) and classical truth (defensibility).

Effective truth is the more fundamental. Whatever is effectively true is also classically true, but not vice versa.

Classical defenses use a further liberalization of the general dialogue rule. This further liberalization also results if we retain the effective dialogue rule but permit additional hypotheses. Here it is the case that a position

$$\Gamma \quad || \quad A$$

is classically defensible precisely if the *effective* dialogue, which takes for every thesis of the proponent that begins with A a corresponding stability hypothesis $A \to A$, is winnable. That is, if an effective dialogue arrives at the position

$$\Gamma \quad || \quad \begin{matrix} A \\ B \end{matrix}$$

and in accordance with the *classical* rule the proponent wants to repeat his thesis A (with which he can subsequently repeat a defense of A), then he may do this *effectively* by adding the stability hypothesis $\daleth A \to A$.

$$\begin{matrix} \Gamma \\ \daleth A \to A \end{matrix} \quad || \quad A$$

The *effective* dialogue looks then like this:

$$
\begin{array}{c|c}
\Gamma & \\
\neg\neg A \to A & A \\
\quad\quad\quad A & \neg\neg A \\
\neg A & A \\
& \vdots \\
& B \\
\end{array}
$$

The right branch is trivially closed. The proponent can now continue the left branch with A, because A is *effectively* allowed as an attack on $\neg A$.

On the basis of this simple example, I think we would be justified in saying that *classical dialogues are only an (occasionally expedient) variant of effective dialogues*. In particular, the classical dialogue rule is always expedient when all the propositions involved are stable, that is, when $\neg\neg A \to A$ is effectively true.

Since Goedel's famous proofs,[8] we have known that all logically composite propositions are stable when they are logically constructed without *disjunctions* from stable primary propositions. A simple reinterpretation—we read $\neg\neg B_1 \wedge \neg B_2$ instead of $B_1 \vee B_2$; $\neg\bigwedge_x \neg B(x)$ instead of $\bigvee_x B(x)$—suffices to make every classically true proposition an effectively true proposition.

Goedel used this reinterpretation to show that the consistency of effective (intuitionistic) logic is just as "unproved" (questionable, dubitable, or however we want to put it) as that of classical logic. Now, however, this reinterpretation produces a trivial consistency proof of classical dialogues in that we can draw upon the consistency proof (via the cut principle) for effective dialogues.

Following these justifications of the general dialogue rules, there remains only one thing to be clarified. In what relation do the material dialogues considered here stand to logical calculi and, thus, to *formal logic*?

There are no "strict" logical calculi that correspond to strict dialogues. This is immediately obvious if we recall how one moves from effective dialogues to an (effective) formal logic. It seems to me that Aristotle must have seen this, although it can scarcely be read from the texts in any com-

---

8. Kurt Goedel, "Ueber formal unentscheidbare Saetze der principia Mathematica und verwandter Systeme I," *Monatschefte fuer Mathematik und Physik* 38 (1931): 175–98.

pelling fashion. He did see, however, that a respondent as opponent who affirmed S a M and M a P "must" also affirm S a P, wholly irrespective of what S and P are—and of whether he had posited at the beginning the thesis ⌐S a P, that is, S o P. It is easy to give a dialogic justification of this necessity.

Ignoring a universal quantifier, S a P has the form of a conditional. If we write s and p for x ε S and x ε P, respectively, then we have the following dialogue position (with m for x ε M):

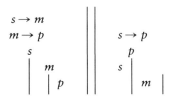

and subsequently the following effective development

$$
\begin{array}{c|c}
s \to m & s \to p \\
m \to p & p \\
s & s \\
\quad | \; m & \quad | \; m \\
\quad\quad | \; p &
\end{array}
$$

In this way, we reach in each of these branches a position

$$
\begin{array}{c||c}
\Gamma & \\
c & c
\end{array}
$$

in which on both right and left there appears the *same* primary proposition. Such a position is called *formally closed*.

Now, what Aristotle must have seen is this: A formally closed position is always victorious. Whether c is true or false, in both cases there is either a true formula to the right of ‖ or a false formula to the left of ‖.

As regards dialogue positions that can be so defended that they are finally formally closed in *all* branches, we say that the hypotheses (as premises) *logically* imply the theses (as conclusions). That is the dialogic definition of logical implication.

If a thesis without hypotheses can be formally closed, then it is called *logically true*.

Having looked first at effective dialogues, we have defined effective logical implication and effective logical truth more precisely. These definitions are not applicable to strict dialogues. Aristotle had also to have seen that the investigation of logical implication and of logical truth is from this point most expediently taken further by using formulas instead of material propositions.

Therefore, we now take $a, b, c, \ldots$ as primary formulas. From these primary formulas further formulas $A, B, \ldots$ are constructed using logical operators. We will no longer speak of material truth or falsity; it will suffice to arrive at formal closedness in all branches.

For example, we can now prove that $\neg\neg a \to a$ is *not* effective-logically true (rather, it is only classically true).

$$
\begin{array}{c|c}
 & \neg\neg a \to a \\
\neg\neg a & a \\
 & \neg a \\
a & \wedge
\end{array}
$$

In this way, we arrive at the position

$$
\begin{array}{c|c}
\neg\neg a & \\
a & \wedge
\end{array}
$$

which is in no way formally closed.

Only if the a of the second row may be repeated (as is permitted in classical dialogues) does a formally closed position result.

When quantifiers occur in material dialogues, constants (e.g., in arithmetic, number symbols) must also be replaced by variables when we make the move to formal logic. In material dialogues the development for theses that are general propositions runs as follows:

$$
\begin{array}{c|c}
\Gamma & \bigwedge_x A(x) \\
 & A(1) \mid A(2)
\end{array}
$$

with potentially infinite branching.

Instead of constants, we use free variables. If the following dialogue is formally closed in all branches,

$$
\begin{array}{c|c}
\Gamma & A(y)
\end{array}
$$

with the variable y occurring in neither Γ nor $\bigwedge_x A(x)$ (i.e., y is "new," as we say), then the potentially infinite branchings in the material dialogue are also formally closed; we need only replace successive occurrences of y by constants. In formal closedness nothing is changed by such a replacement, because this replacement changes neither Γ nor $\bigwedge_x A(x)$—"y" is indeed new.

We have also known since Goedel that this relation does not hold in the other direction. If $\Gamma \parallel A(n)$ is closed for all n, $\Gamma \parallel A(y)$ with a free variable y need not therefore be closed. This should really not have been an unexpected result; nevertheless, Goedel's ω-incompleteness came in fact as a great surprise in 1931.

It is only when formal quantification (using variables instead of constants) replaced material quantification that logical calculi arose. With this development we now have as yet uninterpreted primary formulas

$$a, b, \ldots$$
$$ax, by, \ldots$$
$$axy, \ldots$$
$$\ldots$$

and we no longer have infinite branching.

The development of a given position to the point where it is formally closed in all branches, if possible at all, requires a finite number of steps.

What we now have is precisely the effective (intuitionistic) Beth calculus, which when read upside down becomes a Gentzen calculus. Hilbert calculi are only inexpedient variants of this sequence calculus.

The move from material classical dialogues to classical logical calculi proceeds correspondingly. In this dialogic justification all calculi are proven to be consistent by using the cut principle. Consistency is here defined as material consistency: In every interpretation of the formulas, logical implication always leads from materially true propositions to materially true propositions. This consistency (soundness) is valid for effective dialogues as well as for classical dialogues and rests upon neither intuition nor plausibility but rather upon the dialogic definitions of the logical operators. We demonstrate consistency for the general dialogue rule by using the cut principle. The move to formal logic then uses only the trivialities that (1) formally closed positions are always materially closed and that (2) formal closedness under formal quantification always produces formal closedness under material quantification.

As has become usual since Goedel, all that remains now is to investigate the reverse, so-called *completeness*. When a formula is, for example, effectively true under *every* interpretation, is the formula then also effectively logically true? A remarkable question.

That it nevertheless occurs is easily understandable, if one proceeds using logical calculi of the Hilbert type (as did Goedel). In such a case one is only intuitively convinced that for effective logic the posited "axioms" of logical calculi are materially consistent (correct). In the case of classical propositional logic this material consistency can be demonstrated using truth value interpretation. However, for classical quantification logic, appeal is again made to intuition—to classical intuition; one merely uses classical quantification logic on the metalevel.

Now, however, one arrives at the conclusion that the axioms are "correct" (sound). The obvious question remains: Have sufficient axioms been posited to deduce *all* logical truths from these axioms using the rule of inference of modus ponens? Have, perhaps, one or two axioms been forgotten? Whence the inquiry into completeness.

This inquiry becomes obsolete once we have a dialogic justification of logical calculi. Every logical operator is introduced via a dialogic definition. After the general rules for material dialogues have been justified, formal dialogues with their formal closure are defined as expedient instruments for the investigation of material dialogues.

By definition, logically true formulas are those formulas that can be closed in the Beth calculus. That we lose something of material truth when we go over to merely formal truth shows the incompleteness of arithmetic.

The so-called semantic question of completeness remains remarkable—an example, in the classical case, is a formula that is classically true on *every* interpretation also classically logically true. How, then, do we propose to be able to know that a formula is true on every interpretation? Even if as regards interpretations we restrict ourselves to arithmetic propositions as material propositions, this region of "all" arithmetic propositions is indefinite; no construction of definite propositional regions can exhaust it. How then do we propose to demonstrate anything about it? Now, we demonstrate an assertion concernig "all" propositions precisely by demonstrating that the assertion is logically true. But, in fact, formal logic was introduced only because with it we can demonstrate something about the indefinite region of *all* propositions.

If we follow the dialogic justification, the completeness problem does in fact become obsolete. Sequence calculi are Beth calculi read upside down, and logical calculi of the Hilbert type are shown to be "complete" simply by showing that in them all formulas $B \to A$ are deducible for which $B \prec A$ is deducible in the sequence calculus. With the pragmatics of dialogic justification for Beth calculi, there no longer arises any need for so-called semantics.

At this point I can well imagine that those people who are not familiar with dialogic practice as the basis for all logic are confused by the fact that Goedel nonetheless gave a completeness proof for clasical quantification logic—and that in the last twenty years since Kripke semantic completeness proofs of modal logic and of intuitionistic logic have been all the rage. Therefore, I will try to show how we should understand these completeness proofs when we have ceased to spin illusions about semantics.

Viewed without bias, the Goedel proof indicates nothing concerning "all" arithmetic propositions. Rather, it is based on the nondeducibility of a formula. In the Beth calculus, that is the nondeducibility of a position

$$B \quad \| \quad A$$

In general, the question of closability is not decidable. But we can easily define so-called special developments such that, when a position can be closed at all, closure can also be achieved using a special development.

In some of the cases in which we reach no decision after a finite number of steps we are able to prove that a branch will remain unterminated even after infinitely many steps. In such cases classical metamathematics uses König's lemma to demonstrate the "existence" of an infinite branch, which does not alter the fact that we will be able to construct such a branch in only some cases.

If we have such a branch, then it results almost trivially that the branch supplies a countermodel for the starting position: We take as "true" primary propositions the primary formulas that occur in the branch to the left of $\|$ and as "false" primary propositions the primary formulas that occur in the branch to the right of $\|$. The free variables that appear in the branches are taken to be constants. (For the sake of simplicity we restrict ourselves here to propositions containing no conditionals.) The development rules are then, in this interpretation of the formulas that occur in the branches, precisely those rules under which all formulas to the left of $\|$ become strictly true and all formulas to the right of $\|$ become strictly false. Thus, in particular, B becomes strictly false and A strictly true. Now, then, you can call that "semantics," but it is only a word.

For every classical nonterminatable position there exists (in the classical sense) a strict countermodel.

That is what Goedel actually proved. If we substitute $\neg \wedge \neg$ for the metamathematical quantifier $\vee$, then we obtain from it an effectively true statement. The utility of this is that a special development procedure is given that always leads to a termination, if the starting position is actually terminatable. Now, in my opinion, Kripke's contribution is to have taken this development procedure for classical quantification logic and generalized it to classical modal logic.[9] Via a modal logical interpretation of effective quantification logic, this has also then led to special development procedures for effective quantification logic—and we can always then try to demonstrate the nonterminatability of a position via the construction of a countermodel.

When one proceeds in the usual classical naiveté from the strict dialogue as semantics and then in a circular fashion proves the material (semantic) consistency of classical calculi by using classical logic on the metalevel, then Goedel's completeness principle makes minimal sense in that it takes the countermodel from the strict dialogue.

In the case of Kripke, we find no semantics *prior to* modal calculi; instead, a definition of modal logical truth is chosen only because the definition fits the modal calculi.

Following a tradition set by Lewis, calculi of the Hilbert type are by and large still used for modal calculi. There is added to modus ponens a further rule of inference, usually $A \Rightarrow \Delta A$ (with Δ standing for so-called necessity). The rest then consists of further axioms, in particular,

$$(*) \quad \Delta A \to A$$

and the Lewis axiom

$$(L) \quad \Delta A \to \Delta \Delta A$$

The interpretation of these is left to unconstrained intuition—and it is only in this context that we may understand the acceptance of Kripke's modal theory as "semantics."

In any case, the fact that Aristotle had long ago worked out in detail a modal syllogistic should make us skeptical toward modern fantasy interpretations. Aristotle had, we might suspect, something much simpler in mind when he spoke of Ananke as logical necessity.

If we start with eristic logic, there is a simple route leading to a modal logic. We stipulate that the opponent in every dialogue accepts a certain minimum system of statements: a minimum system Σ of generally accepted knowledge, a priori knowledge, or however you want to put it.

Relative to this knowledge, we can then define the "relative necessity" of propositions by

$$\Delta_\Sigma A \Leftrightarrow \Sigma A$$

---

9. Saul Kripke, "A Completeness Theorem in Modal Logic," *Journal of Symbolic Logic* 24 (1959): 1–15; cf. Kripke, "Semantical Considerations on Modal Logic," *Acta Philosophica Fennica* 16 (1963): 63–72.

By itself, this definition does not lead to modal logic. But Aristotle had already noted that one can easily arrive at implications between these relative necessities that are *uniformly* valid for all Σ. In this way, we obtain a *modal logic*, because these implications on the metalevel are *independent* of the specification of a particular Σ.

We even find in Aristotle (at least I read it in 1.8 of the *Prior Analytics*) the insight that an implication of the form

$$\Delta B_1 \wedge \ldots \wedge \Delta B_n < DA$$

is (uniformly for all Σ) valid, precisely when

$$B_1 \wedge \ldots \wedge B_n < A$$

is valid. For the sake of simplicity Δ is written here instead of Δ_Σ. This trivial metalogical statement clearly justifies the following modal calculus. We add Δ as a logical operator (like negation) to quantification logic. A position consisting of modal formulas

$$\Gamma \quad || \quad A$$

is developed using quantification logic until we arrive at a position

$$\Gamma(\Delta B_1, \ldots, \Delta B_n) \quad || \quad \Delta A$$

As the Δ-step we allow the following development:

$$\frac{\Gamma(\Delta B_1, \ldots, \Delta B_n)}{B_1} \quad \Bigg|\Bigg| \quad \frac{\Delta A}{A}$$
$$\vdots$$
$$B_n$$

The horizontal line signifies that all formulas above the bar can no longer be used for further development. With this justification for a modal calculus there is no longer any reason to continue to seek after a semantics.

The axiom $*$ is trivial, because Σ is presupposed to be true; on the other hand, the Lewis axiom needs only to be investigated as a means for achieving in a consistent fashion a formal simplification in the case of the admissibility of repeated modalities.

None of this is difficult and, as Kripke has shown, the special development procedures of Goedel's completeness proof can be extended to modal calculi. If a modal position

$$B \quad || \quad A$$

is nonterminatable, then we try to construct an open branch. And, finally, we call the sets of primary formulas that occur in that branch a countermodel (due to the Δ-steps, these formulas appear hierarchically ordered).

The existence of such models contributes nothing to a *justification* of modal calculi, because it is only the modal calculi that justify the definition of specific hierarchies of prime formulas as "models." On the other hand, without an appeal to a dialogic interpretation, there are apparently no boundaries set to speculation concerning a semantics of possible worlds.

That we should then take such a semantics for classical logic and use it to "justify" effective quantification logic seems crazy to me. It is not formally difficult, however, because it requires only a simple mapping of effective formulas A onto modal formulas A:

$$\begin{array}{rcl} \overline{p} & \Leftrightarrow & \Delta \; p \text{ for primary formulas} \\ \overline{\daleth A} & \Leftrightarrow & \Delta \overline{\daleth A} \\ \overline{B \to A} & \Leftrightarrow & \Delta(\overline{B} \to \overline{A}) \\ \overline{\bigwedge_x A(x)} & \Leftrightarrow & \Delta \bigwedge_x \overline{A(x)} \end{array}$$

It follows immediately that A is effectively logically true precisely when A is logically true in classical modal logic (with the axioms * and L). This mapping is irrelevant for the problem of justification, because the dialogic justification produces the calculus for effective quantification logic after definition of the logical operators as the first formal logic; everything else must then be justified on that basis.

PART III

Philosophy of Language

SEVEN

Logical Structures in Language

Certain fields of knowledge take natural language (be it one, several, or possibly all natural languages) as their proper object, whereas in contrast other disciplines have only an external relation to language; that is, these latter disciplines take a nonlinguistic object, say, some part of nature or a historical event. In the latter case, workers in these fields to a certain extent speak only out of compulsion, only because they need to communicate their knowledge.

The relation of formal logic to language resembles neither of these possibilities. A logician is not a philologist. On the contrary, he is often a misologist, because he tends to eye natural languages as merely a kind of encroaching weed. But neither can we say that the logician takes a nonlinguistic object as the object of his discipline. Today, people generally accept definitions of formal logic of the following sort: Logic is the science of the truth of propositions on the basis of form alone. According to this definition, logic is clearly somehow related to natural language, for what could propositions be other than linguistic constructions? And it would make no sense here to speak of truth, if what were at issue were not the truth of linguistic propositions. Despite the foregoing observation, in the following I want to demonstrate the extent to which logic is independent of natural language. It is merely that logical structures show themselves in the most perspicuous way in language. The validity of these structures, however, does not rest upon the fact that such structures are embodied in natural language. That is the claim I want to defend here.

The most important part of the definition of logic given above—"the science of the truth of propositions on the basis of form alone"—is the last part, "on the basis of form alone." A simple example suffices to convey what is meant here by "form."

If Socrates is either sitting or standing and if Socrates is not standing, then Socrates is sitting. Anyone with an adequate grasp of English recognizes at once that this proposition is true. In addition, such a person also understands that this proposition is true not because it speaks of Socrates and of sitting and standing but on the basis of form alone. Even were the contents arbitrarily different, that is, if only the form remained preserved (if A or B and if not B, then A), the proposition would be true. As an abbreviation, we say that it is "logically true."

This example shows that discovering something like logical structures in language presents no difficulties. What is a problem is the question: How do these logical structures enter into language? Are they nothing other than syntactic conventions of language, or is their validity based on something external to language? Having asserted the linguistic independence of logic, I am committed to showing that the logical structures that arise in language are not syntactic conventions of language.

What, then, are they? We must return to our example. You can see in the example that the form of a proposition is the way in which the proposition has been constructed using elementary propositions. The letters A and B stand for any given elementary propositions, such as "Socrates is sitting," "Socrates is standing," and so on. The construction is effected by using logical operators, of which the conjunctions "and," "or," "not," "if/then" are examples. The so-called quantifiers, "all" and "some," are also logical operators.

These illustrations of the restrictive specification, "on the basis of form alone," all make use of the English language. In what follows the point is to establish the logical significance of conjunctions (to which I will now restrict my discussion) without thereby having recourse to English syntax. Of course, this does not mean that I will now stop speaking English. Rather, just as multiplication tables written in arabic numerals are taught in all countries independently of the idiosyncracies of the language spoken in each country, the same can be done with logic. Of course, while demonstrating this it will be legitimate for me to make use, for example, of the English language. After all, arithmetic textbooks are also written in English.

We must now examine elementary propositions. We know, of course, that logic is not dependent upon any specific elementary propositions; there is nonetheless no logic without some elementary propositions. As is well known, the simplest structure possessed by linguistic propositions is the subject–predicate structure, as illustrated by our example "Socrates is sitting." In this structure the subject is instantiated by using the proper name "Socrates." The subject is then given a predicate, for example, "is sitting," "is standing," and so on. If you give the predicate "is true" to the proposition "Socrates is sitting," that simply means that you are prepared to give to the subject, in this case Socrates, the predicate "is sitting."

Unfortunately, the assigning of predicates is often a matter of the idiosyncracies of the speaker. In many cases, however, there is an objective decision procedure for ascertaining whether or not the elementary proposition in question is true. I have chosen a nonlinguistic example that employs a dice cup and a thrown die. The die remains covered by the cup. In place of a proposition about the covered die, let someone now place a second die with a specific side up on the table, say, 6. This action represents a proposition: The subject is the covered die; the predicate is the side up on the uncovered die placed on the table. The decision about the truth of the proposition is made through direct personal observation by lifting the dice cup. Only if the first die also shows a 6 is the "proposition" true; otherwise, it is false.

As you can learn in any logic textbook, from the time of Aristotle up to the present logic has made use of the assumption that, somehow or other, a person can decide whether an elementary proposition is true or false. That is, elementary propositions are "in themselves" true or false. We want to call propositions for which this is true *truth-definite*. Although it has so far not been generally accepted, it seems to me that one of the most important discoveries of modern logic is the discovery by the Dutch mathematician Brouwer that in mathematics, where we depend decisively upon logical inference, we are not concerned solely with truth-definite propositions. Let us take an arithmetic proposition:

Some odd numbers are perfect

The notion of a perfect number is a concept that goes back to the Pythagoreans. If a number equals the sum of its proper factors it is said to be "perfect"; for example, $6 = 1 + 2 + 3$ and $28 = 1 + 2 + 4 + 7 + 14$ are perfect numbers. In fact, they are even perfect numbers. No one has thus far found an odd perfect number. Perhaps one will be found in the future—no one knows; hitherto, mathematicians have not been able to exclude this possibility. Thus there is no procedure to decide the truth of the arithmetic proposition above. This proposition is not truth-definite. If a person nonetheless offers this arithmetic proposition as "in itself" true or false, when we do not know it to be so, then that person makes an empty assertion. The positivists ought to label this proposition metaphysical, and yet they do not, because they are, at least as logical positivists, completely uncritical in their acceptance of classical logic. Perhaps they also fear that on their criterion, if they acknowledge this proposition to be not truth-definite, the proposition must be classified as "meaningless."

If someone were to assert this arithmetic proposition, you would not answer him; you would not understand him. In any event, it would make more sense to ask him whether he could prove his proposition. What does that mean in this case? Now, any attempt to prove this proposition must look like this: An odd number n must be chosen, and we must then decide whether this number n is perfect. Every proposition of the form "n is perfect" is truth-definite, because we need only calculate the sum of its proper factors and compare the sum to n. If you choose an n that results in an affirmative decision, then the proof is successful; otherwise, it fails. A successful proof is called simply a *proof* for short.

We have here a procedure that decides for every attempted proof whether or not it is a proof. In such cases we will call the proposition *proof-definite*. For "provability" we can also give a nonlinguistic example. Think of children who bet on diving into a swimming pool for stones. Whoever throws a stone into the water, perhaps particularly far, proposes or bets through this action that he can retrieve the stone. His action then represents a proposition that is proof-definite in that his attempted proof is successful only if he jumps into the water and surfaces with the stone.

In general, the situation can be described as follows. An action, for example, the assertion of a proposition, is proof-definite if a procedure is given that confers on a further action—that is, the attempted proof—exactly one of two possible values ("successful" or "unsuccessful").

As regards the elementary propositions $a, b, \ldots$, from which we will construct further assertions using logical operators, we now want to assume only that they are proof-definite in this sense. If we connect the proof-definite propositions a and b using "and," we need to define what meaning "a and b" is to have. Which procedure, performed as an attempted proof, should be described as successful? If we take our cue from the meaning of "and" in natural language, we will say that a proof of a proposition of this form results from two procedures, a proof of a and a proof of b. In this way, "a and b" becomes a proof-definite proposition. It is just as easy to define "a or b" as a proof-definite proposition: If you want to prove "a or b," you must offer either a proof of a or a proof of b, whichever you prefer. I call this way of interpretating the meaning of conjunctions their operational interpretation, because it specifies their meaning by using schematic procedures.

This interpretation differs from the customary truth-table interpretation of classical logic, as can be seen clearly in the case of negation. The meaning of "not" in natural language does not supply us with a way to establish when an attempted proof of "not-a" succeeds or fails in cases where a is proof-definite. To illustrate this, let me introduce the situation of two people engaged in a dialogue. If one person asserts a proof-definite proposition, the other person can ask him for a proof. Only when the attempted proof succeeds, for which we need a decision procedure, is the dialogue won rather than lost. The assertion "not-a" can be employed meaningfully in such a dialogue. That is, the participants can agree to the following: If "not-a" is asserted by one party, the other of course may not demand a proof of this proposition but may instead assert a. If he can defend the assertion of a, the first party has lost rather than won. Of course, by using this procedure "not-a" does not become a proof-definite proposition, but there is specified a procedure for the employment of this proposition in a dialogue—a procedure that decides "success" or "failure" within the dialogue.

We will call every proposition for which such a procedure has been specified *dialogue-definite*. In this way, negation is also operationally interpreted for proof-definite propositions. The parties to the dialogue know what they have to do in order to carry on the dialogue.

We could also formulate this operational interpretation in the following way. If a person asserts "not-a," that person has agreed that, if the other person can defend the proposition a, then he will have lost. The "if/then" that occurs here is a conjunction that is used in natural language and that the classical logic of connectives has been notoriously unable to handle adequately. On the other hand, using this operational interpretation, every if/then conjunction of dialogue-definite propositions can now be interpreted easily as a dialogue-definite proposition. If one party to a dialogue asserts "if a, then b," then the other party can assert a. If he can defend his assertion of a, then the first party must assert and defend b.

I should give an example of how such a dialogue would run. We record the propositions asserted by both parties in a tableau. On the right we record the propositions asserted by the first party. We will call the first party the proponent, P. On the left, then, we will record the propositions asserted by the opponent, O.

Opponent	Proponent
	If a or b and not b, then a
a or b and not b	?1
a or b	?
a | b	a
not b	
	?2
	b

The dialogue ends with the victory of the proponent, *P*. The tableau even shows that *P* can always win the dialogue, independently of the content of the assertions *a* and *b*. Thus the tableau shows a winning strategy on the basis of form alone. So we may call the assertion "if *a* or *b* and not *b*, then *a*" logically true. However, this logical truth is distinguished from the logical truth of classical logic. Brouwer gave this new logic the name of intuitionistic logic, although the phrase is not exactly a fortunate choice.

I will give here yet another example. Is the proposition "*if not a and b, then not a or not b*" logically true in an intuitionistic sense? In classical logic "*not a and b*" *implies* "*not a or not b*." Nonetheless, our proposition is not logically true in an intuitionistic sense. This can be seen clearly in the course of the dialogue.

Opponent			Proponent		
			If not *a* and *b*, then *not a* or not *b*		
not *a* and *b*			*a* and *b*		not *a* or not *b*
?1	?2	?	*a*	*b*	not *a*	not *b*
?	?	*a*|*b*			?	?

P could win the dialogue only if he (1) knew how *a* and *b* can be defended or (2) which one of the two assertions by *O* cannot be defended. Of course, *P* may well know these things for specific propositions *a* and *b*, but then he would know these things only on the basis of the content of *a* and *b*. Under no circumstances can he win the dialogue on the basis of form alone. Thus his assertion is not logically true in an intuitionistic sense.

It is somewhat misleading to formulate all this in such a way that one speaks of different logics. The word "logic" is much better employed in the singular only. Every proposition that is logically true in an intuitionistic sense is also logically true in the classical sense. This is obviously so, because truth-definite propositions are nothing more than a special case of dialogue-definite propositions. The reverse is, however, not the case. Classical logic is a special case of intuitionistic logic, and the latter is, so to speak, the true logic, for it is valid not only for truth-definite propositions.

The arithmetic example, "Some odd numbers are perfect," has also shown that intuitionistic logic remains applicable in cases where the quantifiers "some" and "all" of Aristotelian logic are employed externally to the connectives in the construction of propositions. The example shows that the existential quantifier "some" can be interpreted in terms of provability. To defend the proposition "for some x: $a(x)$" you must be able to name an object n from the domain of the variable x and then be able to defend the proposition $a(n)$. Thus, in the case of "some," the proponent P himself chooses the object n. On the other hand, the universal quantifier "all" must be interpreted in terms of dialogue-definiteness. To defend the proposition "for every x: $a(x)$" you must be able to defend the proposition $a(n)$ where the opponent can choose whatever object n he wishes.

The definition of intuitionistic logical truth remains the same: A proposition is logically true, if it can be successfully defended on the basis of form alone in dialogue against every possible opponent. If a proposition is logically true in an intuitionistic sense, it must always be possible to outline a winning strategy for it. Using the operational interpretation of logical operators allows us systematically to map out the winning strategy. The rules of construction for the production of all possible winning strategies can then be given in the form of a tableau as was done above. A winning strategy always consists of its being possible for a certain starting position of dialogue to be traced back step by step to certain trivial concluding positions, namely, those positions in which the proponent needs only to put forward one elementary proposition, one that has previously been put forward by the opponent. If the tableau for such a winning strategy is read from bottom to top, we obtain a procedure for the construction of all logically true propositional forms, a procedure that begins with assertions that are trivially logically true and that can thus function as logical axioms. In this way, we obtain nothing other than precisely what has been long known in the literature as a logical calculus. In particular, we obtain Gentzen's intuitionistic sequence calculus. However, this logical calculus proves here to be only an easy ancillary method for investigating winning strategies. A person who has mastered these winning strategies must be said to think logically, even if that person has never heard a thing about logical calculi. Of course, we learn this ability to think logically just as we learn a language or as we learn arithmetic calculation with numerals. But logical thinking itself is nonetheless no more bound by natural language than is arithmetic. The dialogues that we have recorded in tableau form represent logical structures that are of course first encountered in language but that can also be instantiated in nonlinguistic action.

The essence of the dialogue situation is also found in competitive activities. If we think back to the example of diving for stones, we can imagine that in order to make this into a contest a player could throw two stones into the water and then signal, say, by a simple gesture like the lifting of one arm or both arms, that he was committing himself to retrieve both or at least one of the stones. Or, with another gesture, he could indicate that he was committing himself to retrieve the second, if the first were retrieved by someone else. Such games can also be played as card games. Elementary propositions would then be represented by playing cards, so that the variables in our tableau would be nothing other than chips. The operational interpretations of connectives, for which some cards must be present, then determine the moves in the game.

If we want to understand the logical structures in natural language, it seems to me indispensable that we be able to instantiate such logical structures nonlinguistically. That is precisely what is accomplished with operational interpretations of the logical operators. At the same time, this shows both in what sense logic is independent of natural language and how logical structures can enter into language.

EIGHT

Logic and Grammar

In the preface to the Duden German Grammar you will find the following statement: "In an earlier era the assumption was that one could look only for certain universally valid structural forms (of language) in the German language."

In contrast, the Duden Grammar begins with the insight that each individual language has its own structure, which is "the result of the language community's linguistic grasp of the realities of the world."

The fortunate effect of this general insight is that there are now available German grammars, as well as English and others, that do not borrow their grammatical categories from Latin grammar. The fact that Western culture is a derivative culture is nowhere more obvious than in the fact that it is only in this century that European languages have consciously achieved their grammatical independence.

To avoid misunderstanding, I want at the outset to emphasize that when in speaking of logic I speak of universally valid structural forms of language I have no intention of subverting this autonomy so hard-won by our Western European mother tongues. Against the more radical inclinations of "ethnolinguists," who claim that there is no such thing as "thought simpliciter" and, rather, only a spectrum of linguistically bound thought, I only want to point out the possibility of a logical use of language, as I call it, for which the specific forms of individual natural language are *irrelevant*.

I will not try to talk you into believing that such possibilities are possible. Rather, I want to place before your eyes several examples of the simplest of these possibilities. In doing this I will, of course, continue to speak English. I ask, however, that you take note that the peculiarities of English are irrelevant to what I shall say. If this appears doubtful to you in any of the examples, we need to discuss the example further. Once again, my thesis is not that *all* English grammatical structures should be understood as instantiations of logical structures. Rather, my thesis is that in addition to the idiosyncratic structures of English there are also *some* structures that are better understood as instantiations of logical structures. This has absolutely nothing to do with a "structuralist" conception of language and equally little to do with a "logicistic" conception. For me it is not a matter of programmatically propounding a direction for research. For me it is merely that some structures in English are better understood by viewing them as instantiations of logical structures than by ignoring or denying the existence of such a logical standard. Under no circumstances do I want to occasion a fight over the importance of the logical structures compared to the idiosyncratic structures. I would much rather set for myself the task of demonstrating that, be that as it may, there are also logical structures in English.

I will restrict myself to four examples of these logical structures, which I call:

>elementary propositions
>terms
>concepts
>logical operators.

1. ELEMENTARY PROPOSITIONS

To anyone who tries to analyze it, natural language at first appears to be a totally unsurveyable complex. Nonetheless, it is possible to select from that portion of a natural language that comprises sentences in propositional form certain elementary sentences. These elementary sentences are characterized by our not being able to break them down into any simpler propositions. In English these elementary propositions have the form "This is so" and "This is not so." The oldest example of an elementary proposition is found in Plato, "Theaetetus flies." Other examples are, for example, "Khrushchev is no philosopher" or "Faust loves Gretchen." You can see that in an elementary proposition there first appears a *proper name*—Theaetetus, Khrushchev, Faust, and so on—which can be *historical, actual,* or *poetical* proper names. These can, of course, be the proper names not only of persons, as in these examples, but of any given object. The second significant constituent part of an elementary proposition I would like to call the *predicator.* Predicators are words that are either attributed to or denied of an object. In this connection "predicator" indicates a kind of word and not a sentential part like the grammatical term "predicate." In all languages objects are designated by proper names, and predicators are either attributed to or denied of objects. On the other hand, the way in which the attribution or denial is expressed varies from language to language. In English it sometimes occurs with a copula, "is," and sometimes without the copula. In the following I will represent the logical form of elementary propositions by beginning with the relevant proper name, followed by the copula, and finishing with the predicator. For the copula I want to use the Greek letter ε, from ἐστί, for attribution (affirmation) and ε' for denial (negation). If we then use, say, capital letters as abbreviations of proper names and small letters as abbreviations of predicators, we get the following logical form for our examples:

>Theaetetus flies: $T \, \varepsilon \, f$
>Khrushchev is no philosopher: $K \, \varepsilon' \, p$
>Faust loves Gretchen: $F, G \, \varepsilon \, l$

The last example illustrates a two-valued predicator. It is logically irrelevant that in grammar the first proper name is designated the subject and the second is designted the object. In addition to one-valued and two-valued predicators, there are also many-valued predicators—in theory, any given number. In practice, however, there rarely occur more than three-valued or four-valued predicators.

2. TERMS

Suppose we restrict ourselves to using only elementary propositions. Then, when there arose a difference of opinion over whether a certain predicator was to be attributed to or denied of an object, we would only have recourse to adducing examples or counterexamples for the intended usage of the predicator. We could say that the predicator p should in every case be attributed to certain objects, $O_1, O_2, \ldots$, and that the predicator p should in every case be denied of certain other objects, $H_1, H_2, \ldots$ If we have only elementary propositions, predicators are only, as I put it, *exemplarily specified*. In such exemplary specifications the usage of the predicator is obviously not specified very completely. If we wish to fix the usage of predicators more exactly, we have the possibility of introducing *rules* for the employment of predicators. The simplest of such rules have been given to us by Aristotle, who made them the foundation of his logic, the so-called syllogistic. Aristotle restricted himself to one-valued predicators and gave two kinds of rules: rules of affirmation that allow the transition from a proposition "$X \, \varepsilon \, p$" to an affirmative proposition "$X \, \varepsilon \, q$" and rules of negation that allow the transition from a proposition "$X \, \varepsilon \, p$" to a negative proposition "$X \, \varepsilon' \, q$."

We render both kinds of rules symbolically as follows:

$$X \varepsilon p \Rightarrow X \varepsilon q$$
$$X \varepsilon p \Rightarrow X \varepsilon' q$$

Here, p and q are two predicators, the usages of which are fixed by one of these rules; X is a variable for any given proper name; and the double arrow, $\Rightarrow$, symbolizes the admissibility of the transition from the left-hand elementary proposition to the right-hand proposition. Individual natural languages contain no such rules in explicit form, because these languages have developed historically. A given linguistic usage in such languages has, therefore, always to be first interpreted as to whether or not a rule for the usage of predicators is intended in it. When a person quotes the following sentence, for example, as a proverb, "Whoever is rational is just," then we can interpret this as the speaker's wanting to prescribe the transition from an elementary proposition of the form "X is rational" to an elementary proposition of the form "X is just." In the same way, we can interpret the following sentences,

>Whoever is just is not compassionate
>What is red is not green

as negative rules that serve to fix the usage of the predicators "just" and "compassionate" and "red" and "green," respectively.

In the following I want to call a predicator that has had its usage fixed by such rules a *term*. We could then also call rules that fix the usage of predicators *terminological specifications*. From the perspective of logic, all terminological specifications are only suggestions for the usage of predicators. The terminological specifications contained implicitly in one's native language are customary traditional suggestions. Although these customary suggestions are indeed necessary for an initial understanding, in specialized discourse we can loose ourselves of them and eventually replace them with others.

Terminological specifications always concern several predicators, at least two, but more generally they concern an entire system of predicators, for example, the system of color predicators. When a system of predicators becomes a system of terms through application of rules that fix their usage, such a system is called a *terminology*.

3. CONCEPTS

Using terms according to a system of rules is often also referred to as conceptual thinking. Thus many people also speak of a *conceptual system* instead of a terminology. The whole history of logic, however, says that "term" does not mean the same thing as "concept." Rather, the distinction between concept and term is that to get concepts we *abstract* from the phonetic form of terms. What exactly we are to understand by such an abstraction has been a constant subject of discussion in philosophy since Aristotle. In scholasticism the fight over universals also centered on the question of the difficulty of this abstraction, and in the contemporary discussion of the foundations of mathematics this problem has become acute once again. In my opinion the difficulty lies in the fact that the Aristotelian texts tend to view concepts as something mental as opposed to something linguistic. The nominalistic solution to this difficulty, which you can accept without having to accept the rest of nominalistic philosophy, lies in taking talk of abstract objects as a *façon de parler*, as a device for expanding the possibilities of speech.

To put this psychologically, at the bottom lies the phenomenon that we can mean the same thing with different words (terms). Consider the following example drawn from the circumstances we are discussing. I have, for example, spoken of our being able to fix the usage of predicators by using rules. If I had instead spoken of our being able to make our employment of words more precise by following precepts, it seems to me that you would have immediately felt that the same thing was being discussed. In any event, it is pointless, for example, to seek an answer to the question, "Do the words 'fix' and 'make precise' *mean* the same thing here or do they not?" We cannot expect an answer as long as we simply take such words out of natural language without expressly specifying their usage more closely. Only when we compare two different terminologies and ascertain that, ignoring phonetic differences, the two systems of predicators have the same exemplary specifications and that the same rules are operant have we established a precise and comprehensible *equivalence* between different terms. Thus we want to speak of concepts rather than terms only when the phonetic form of terms is considered irrelevant (when we abstract from the phonetic form) but the exemplary specifications and the terminological specifications are retained. What, then, is the distinction between the term "red" and the concept "red"? If we call the concept the *reference* of the term, we are using a misleading analogy to proper names that have objects as their reference, namely, objects that are called by the appropriate proper names. If, on the other hand, we call the concept the *content* of the term, we are using a spatial metaphor similar to "the contents of consciousness." Concept understood as that which the term "means" is no longer metaphorical but remains psychological. The act of abstraction that leads from term to concept, however, is not a psychical operation but, rather, a logical one. Instead of throwing ourselves into the ontological difficulties of the question "What is a concept?" we ask ourselves how propositions concerning concepts are constructed. Propositions dealing with terms are not a problem. For example, we can say that the term "red" consists of three letters. In constructing this proposition you cannot abstract from the phonetic form of the term "red." If we had another word instead of "red" that had the same exemplary and terminological specifications, the number of letters might be different. This proposition concerning the term "red" is not invariant with regard to substitution of the term by a different but equivalent term taken from a different terminology. There are, however, propositions that do possess this *invariance;* that is, if the proposition is valid for one term, then it is also valid for all equivalent terms. As an example, consider that the terms "rational" and "compassionate" are *contraries* within the terminology above; that is, no object may have both predicates attributed to it. The fact that the two terms are contraries depends only on the given terminological specifications, not on the phonetic form of the predicators, and is therefore invariant. The abstraction involved consists only in the fact that we restrict ourselves to invariant propositions about terms. In order to give expression to this invariance we then say, "The concepts rational and compassionate are contraries," rather than saying that the terms "rational" and "compassionate" are contraries.

Propositions concerning concepts do not appear in spontaneous, naive speech. We encounter such propositions in reflective speech and particularly in the speech of grammarians. A grammarian must talk about words; he speaks a metalanguage, as people say these days. Talk of concepts serves as a linguistic device that steers the reflection of grammarians toward exemplary and terminological specifications and away from mere phonetic forms. In contrast to phonological linguistics, content-oriented linguistics operates on the level of reflection in order to determine the variety of concepts and the differences in intellectual grasp of individual languages.

4. LOGICAL OPERATORS

Up to this point our observations have not yet led us out of the logical structure of elementary propositions. The terminological specification discussed in sections 2 and 3 were rules, not propositions. When we turn to the use of logical operators to combine elementary propositions, however, we encounter logical structures that are as complicated as one could want. The theory of logical operators rests on Aristotle's work in logic but even more so on the logical work found in the Megarian and the Stoic philosophies. Logical operators are *words used to combine* given propositions, combinatory words that have a fixed usage. Recall that propositions are used in dialogue; a person *asserts* a proposition in conversation with another person. For example, if $A, B, \ldots$ are propositions to which the participants in a dialogue agree, when one participant asserts "A and B," this is to be understood to mean that at the request of the other participant, called the opponent, both A and B must be asserted. On the other hand, if "A or B" is asserted, then the person who has made the assertion, called the proponent, may choose which of the two propositions he will assert at the request of the opponent. The English word "or" is not always used in this way, and for that reason its use in dialogue must be expressly determined and fixed. The functions of the two participants in a dialogue become clearer in the case of the logical operator "if/then." If the proponent asserts a proposition of the form "if A, then B," the opponent has the right to assert the if-clause, "A." If the opponent can defend this assertion against the proponent, then the proponent must assert and defend the then-clause, "B." Negation represents a special case. If the proponent asserts "not A," then the opponent has a right to assert "A." If the opponent can defend his assertion, then the proponent loses the dialogue.

These logical operators, "and," "or," "if/then," and "not," comprise together all the *junctors,* that is, logical operators that combine propositions into more complex propositions. The peculiarities of the English words chosen here to represent the junctors are irrelevant for determining and fixing the dialogic usage of the proposition that result from the use of the junctors. Just as was the case with terminological specifications, in the case of junctors historically developed natural languages contain no explicit rules for the dialogic employment of logical operators. The logical usage of these words (e.g., that we deduce "B" from "A or B" and "not A") cannot simply be read off linguistic usage. Rather, it is logical usage that first provides a standard against which actual linguistic usage can be measured and understood.

Besides the junctors there are also logical operators that are quantificators, or *quantifiers,* for short. There are exactly two of these quantifiers: the universal quantifier "for all" and the existential quantifier "for some." It is only by using these quantifiers that we can use elementary propositions to construct propositions that contain no proper names. Using the propositional form "X is p" we can construct "for all X, X is p" (e.g., "everything is transitory") or "for some X, X is p" (e.g., "something is eternal"). If a universal sentence is asserted by the proponent, then the opponent may choose a given X; if an existential sentence is asserted, then the proponent must himself produce an X. An example would be a dialogue concerning the following propositions: "All atheists are feebleminded or wicked." The logical form of this sentence is:

For all X, if X is a, then X is f or X is w

The opponent is allowed to choose a specific X, and he chooses, for example, "Russell." The dialogue will run somewhat as follows:

Opponent	*Proponent*
Russell?	If R is a, then R is f or R is w
R is a	?
...	R is f or R is w
?	R is w
?	...

(At the points where ". . ." is shown, there would occur a dialogue concerning the elementary proposition that is queried by "?")

The logical operators demonstrate, in particular, that up to the present time grammar, when it ignores logical structures, does not sufficiently and adequately explicate linguistic forms. This applies as well to grammars derived from Latin. In grammar the logical operators are not introduced as a special kind of word: "Not" is presented as an adverb; "and," "or," and "if/then" are presented as conjunctions; and the quantifiers "all" and "some" are presented as indefinite pronouns.

Precisely at the time that Latin, the scholarly language, was replaced by vernacular English, the double negative was no longer used as a strengthened negative, as it is still used in folk dialects. This development in English is easily explained by the fact that the *logical* usage of "not not-A" is the contradiction of "not A." Without reference to the logical usage of language such developments cannot be rendered intelligible.

The examples adduced here do not exhaust logical structures. Modern logic also contains, for example, a "theory" of definite articles. But the number of logical structures remains slight in comparison to the abundance of structures found in natural languages. Therefore, no logic can render the empirical work of grammarians superfluous—but no grammar can absolve the grammarian of the effort to discipline logically his own usage of language.

NINE

Logic and Hermeneutics

The word "logic" denotes the art of thinking; thus it also denotes the theory of correct thinking. Similarly, hermeneutics denotes the art of understanding and also the theory of correct understanding. The close connection between logic and hermeneutics is due to the fact that thinking must precede the acts of speaking and writing sentences—or at least ought to precede them—and that understanding ought to result from the acts of reading and hearing sentences.

By way of initially engaging your interest in these matters, I first want to discuss the present political implications of these theories. To that end I will begin with the East–West conflict. East and West can be characterized in the following way. In the West differences of opinion are resolved by open discussion, whereas in the East such differences are resolved by decrees of the party. The Eastern system compels a unitary public opinion, whereas the Western system of open discussion in contrast has led to a multiplicity of publicly held and represented opinions on all of the so-called principal questions of the day. These principles are a matter of general propositions, that is, propositions that state no particulars, no here and now. Not every general proposition is a principle, but principal questions are always a matter of general propositions. The Western situation with its multiplicity of publicly represented principal opinions is called "pluralism." A more precise term would be "polydoxy," a multiplicity of opinion. I have the impression that people in the West are frequently proud of this pluralism, this polydoxy, because it is proof of the freedom to be found in the West.

It is indeed that. But this pride in pluralism leaves the following presupposition unstated. A multiplicity of publicly held opinions is a necessary consequence of open discussion. It could also be claimed that unity of opinion, monodoxy, can only be compelled. Before we content ourselves with multiplicity of opinion, which is often nothing more than chaos of opinion, before we are even proud of this situation, I believe we should investigate more closely whether it is not the case that besides compelled monodoxy (unity without freedom) and free polydoxy (freedom without unity) it is possible to achieve a free monodoxy (freedom with unity). By the latter I mean a free agreement on all that thought and understanding have brought forward in our so-called principal questions. On specific concrete questions—for example, whether a bridge should be built over a canal or a tunnel dug under it—free agreement is seldom achievable, for in such cases interests must be weighed against other interests. But the discussion of specific questions—for example, discussion of new marriage legislation—must frequently be carried out using arguments that are derived from principles; in our pluralistic society a person argues from a Marxist, positivist, Christian, or some other standpoint.

Consideration and examination of thought and understanding, which is to say consideration of logic and hermeneutics, cannot solve any concrete questions. Such examination could, however, facilitate the discussion of specific questions, if agreement could be reached on the general assertions that are involved in the discussion of specific questions. Such an agreement might be based on a determination that common knowledge supports none of the relevant general assertions.

I must now turn to details so that the possibility of free and principled agreement that I have sketched here will not remain a mere possibility and so that you will not think that I am merely playing here with platitudes and clichés. What does "to offer an opinion" really mean? It means to assert a sentence formulated in a natural language. Sentences that can be asserted are called propositions, in distinction from, say, questions or commands, which are certainly spoken but cannot be asserted. The propositions that we assert belong to a natural language, to our native tongue, as is sometimes said. At this point, however, we encounter a distinction of fundamental importance. Within natural language we distinguish between a part that is said to be practical and a part that is said to be theoretical. The practical part of natural language is a tool of understanding that is used in our ordinary, everyday activities. It is often called ordinary language. The theoretical part, on the other hand, is a result of the refinement of practical assertions found in technical, scientific languages, including philosophy and theology. Naturally, there is a gradual transition from the practical part to the theoretical part, so that there is no sharp boundary to be drawn here. That is not the issue here, however. The thesis I want to consider here can be formulated in the following way: *The refinement of practical assertions into theoretical propositions is a process that can be taught in a methodical fashion*—that is, a process that is analyzable into individual steps that can be taught one after another.

I will try to establish this claim by leading you through some of these steps, one after another. I begin with practical assertions. I can restrict myself here to so-called elementary propositions—that is, propositions of the form "this is so" and "this is not so." An example with which Plato was familiar is "Theaetetus flies." Other examples are "Caesar is no philosopher" and "Faust loves Gretchen." We can write these propositions schematically as follows: T is f; C is not p; F, G is l.

In these elementary propositions there first appears a proper name (which can be historical, contemporary, or poetical) and then a predicator, a word that functions as a predicate in the proposition. Among predicators we can distinguish one-valued, two-valued, and many-valued predicators. As long as we use only elementary propositions, predicators can only be "exemplarily specified" by their use in such elementary propositions. It is a fact that children learn their first predicators only by example; a certain predicator is assigned to certain objects and is withheld from certain other objects. Predicators are the way in which we draw distinctions within the world of objects. The world first receives an articulation through predicators. All elementary propositions are sentences that concern only particulars. We first encounter general sentences when we switch to the possibility of speaking theoretically. I will restrict myself here to four basic steps that lead to theoretical propositions and that I call:

1. terms
2. logical operators
3. definitions
4. descriptions

Let us first consider terms. If there arises a difference of opinion involving practical assertions, the first step that can be taken—and one that I highly recommend—is to clarify the use of the predicators involved by giving as many examples as possible. A second step, which leads to something new, is to fix the usage of the predicators with rules. Rules determine the relationship between two or more predicators. The simplest rule that produces a relation between two predicators, p and q, can be found in the works of Plato and Aristotle and takes the form of either

"x is p" warrants "x is q"

or

"x is p" warrants "x is not q"

This kind of rule articulates operations that can be performed using elementary propositions. That is, in formulating a rule we speak about language; the word "warrant" belongs to the metalanguage, as we say. In ordinary language we formulate such elementary rules, for example, in the following way:

> Whoever is rational is good
> Whoever is just is not compassionate

Just by looking at these sentences we cannot tell whether they are meant as rules. But we can understand them as rules, as rules that are supposed to establish the usage of the predicators involved—in this case, the predicators "rational" and "good" and "just" and "compassionate," respectively.

If a predicator's usage has been set by a valid set of rules, it is then called a *term*. Terms never occur in isolation; they always form a system and are said to belong to a *terminology*. The rules for the use of terms are therefore called *terminological specifications*. Terminological specifications are frequently formulated as universal propositions. But, for example, the sentence "All swans are white" as usually understood is not a terminological specification of "swan" and "white." This kind of sentence is an empirical proposition (and, in fact, a false one).

Terminological specifications are always only suggestions for the use of predicators. Not every suggestion need be taken. Rather, we should want to investigate the extent to which any suggested terminological specifications are appropriate and adequate to the objects that we want to discuss. Natural languages contain many terminological specifications—sometimes quite clearly but more often only rather obscurely. These represent customary and traditional specifications that must also be subjected to critique.

The activity of operating with terms according to a system of rules is also often called *conceptual thinking,* and we therefore often speak of a system of concepts instead of a terminology. The distinction between a concept and a term is that, when we have abstracted from the morphological (verbal and visual) properties of terms and have left only the system of rules, we speak of concepts rather than of terms.

New possibilities for constructing concepts arise when we now move on to logical operators. The theory of logical operators can be traced back to Aristotle and Megarian and Stoic philosophy. Logical operators are words that join together propositions and that do this in a specified manner. Every assertion occurs within a dialogue; that is, one person asserts a proposition to another person. For example, if $A, B, \ldots$ are the kind of propositions upon which the parties to a discussion are able to come to some agreement, then, if one person makes the assertion "*A* and *B*," another person (called the opponent) may demand that both *A* and *B* be asserted individually. If, on the other hand, "*A* or *B*" were asserted, the person making the assertion (called the proponent) may himself choose which of the two propositions he wants to assert. The English word "or" does not always have this meaning, so its use in the dialogue must be specified. The roles of the parties to a discussion become clearer in the case of the logical operator "if/then." Should the proponent assert a proposition of the form "if *A*, then *B*," the opponent may then assert the if-clause, "*A*." If the opponent is able to defend *A* against the proponent, the proponent must then assert and defend the then-clause, "*B*." Negation represents a special case. If the proponent asserts "not *A*," the opponent may assert *A*. If the opponent is able to defend *A*, the proponent loses the dialogue.

With these logical operators ("and," "or," "if/then," and "not") we have essentially all junctors (i.e., logical operators) that serve to join propositions together. In addition to junctors there are also quantificators—quantifiers, for short. There are exactly two quantifiers: the universal quantifier "for all" and the existential quantifier "for some." It is only by using these quantifiers that we can construct elementary propositions that do not contain proper names. Using the propositional form "*X* is *p*" we get "for all *X*, *X* is *p*" (e.g., "everything is mortal") or "for some *X*, *X* is *p*" (e.g., "there is something that is mortal"). If a universal proposition is asserted, the opponent may then choose a given *X*; if an existential proposition is asserted, the proponent must himself produce an *X*. For example, consider a dialogue concerning the following proposition: "All atheists are either feebleminded or evil-minded." The logical form of this proposition is

For all *X*, if *X* is *a*, then *X* is *f* or *X* is *e*

The opponent may now choose a given *X* and chooses, for example, "Russell." The ensuing dialogue might then run as follows:

Opponent	Proponent
Russell?	If *R* is *a*, then *R* is *f* or *R* is *e*
R is *a*	?
...	*R* is *f* or *R* is *e*
?	*R* is *e*
?	...

With the help of the universal quantifier, as already suggested, terminological specifications can be formulated as universal propositions. In addition, however, we also have universal propositions that are empirical; for example, "All cultures have myths."

In particular, the logical operators allow us to move from terminological specifications and from accepted universal empirical propositions or particular propositions to the logical level of deductions from other propositions, $A, A, \ldots$, when the proposition "if A and A and $\ldots$, then B" is logically true. "Logically true" means here that the proposition can be defended in a dialogue solely on the basis of its form, that is, the way the proposition was constructed out of other propositions, using logical operations. In this case the opponent either must have already asserted B in the dialogue or must assert A and thereby lose because he must assert "not A" at the same time.

Logical operators also make possible terminological specifications more complicated than previously considered. Definitions, in which a proposition with a new predicator is introduced as an abbreviation for a complex proposition, represent a special case. Consider the following example drawn from the law. According to the legal statutes, "majority" is defined as follows: "Majority applies to anyone who is at least 21 years old or who is at least 18 years old and has been declared to be in his majority by a judicial proceeding." Although definitions are very important, they are nonetheless only a special case of conceptual specification. The customary and traditional specification in English, "What is red is not green," for example, is not a definition of these color predicators and is moreover also not a logical deduction from some possible definitions.

The final basic step in the construction of theoretical discourse that we will consider is that of descriptions, which can be used in elementary propositions in place of proper names. For example, when you read in the newspaper the headline, "The President of France to Visit," this is equivalent to "De Gaulle to Visit." That is, there is precisely one president of France and that person is at that moment named "de Gaulle." On the other hand, "The King of France to Visit" would not be intelligible, because no one is the king of France. Were you to read "French Minister to Visit," this would also be unintelligible, because there is more than one Franch minister. A group of words that begins with the definite article and is followed by a predicator (possibly a relative clause like "he who is without sin") is called a potential description. A potential description of the form "the p" becomes an actual description in cases where (1) there is something that is p and (2) there is not more than one thing that is p. If both these conditions are not satisfied, then we are dealing with a pseudodescription. Whether or not a potential description is a pseudodescription depends naturally on the prior specifications of the relevant predicators. For example, if, as is the case in Tillich's theology, "God denotes the deepest ground of being," then we will have to ask whether exactly one thing and not more than one thing can be a deepest ground of being. Because Tillich does not explicitly address this issue, we are permitted to assume that Tillich is here appealing to a certain tradition, for example, the Platonic linguistic usage of "being," "higher being," and "the highest being." Only a critical investigation of Plato will be able to decide whether or not Tillich is using a pseudodescription.

These considerations bring us to more powerful methods of theoretical linguistic usage, to hermeneutics. The elementary tools considered to this point are of course not adequate for mathematics and the exact sciences. The fundamental propositions of exact theories (e.g., arithmetic or geometry) are neither terminological specifications nor universal empirical propositions. But here I do not want to take up this problematic, which Kant dealt with under the heading of "synthetic a priori judgments." There is another problem that makes theoretical use of language even more complicated in the human sciences: How can what one person says, particularly what is said in writing, be taken up into our own theoretical thinking? This taking up into our own thinking—that is, the critical appropriation of traditional texts—is the special task of hermeneutics. Here I will raise only two points. A traditional text speaks to us like an unknown and single-minded teacher. That is, first, the linguistic usage of the author is unknown to us, and, second, the author speaks to us only about his own problems, not about our problems. As an example, I will consider two citations from Aristotle. Suppose we would like to know something more precise about the relation between words and concepts, and suppose that we read the following in Book 1 of Aristotle's text concerning propositions, *On Interpretation:* "Spoken words are the symbols of mental experience, and written words are the symbols of spoken words." Anyone who reads a written word knows for which spoken word it stands. Aristotle asserts here that anyone who understands a spoken word would know for which mental experience it stands. Unfortunately, in this context I can only caution against a person's appropriating this statement, for "the mental experiences that are denoted by a specific word" is a pseudodescription. To be sure, this error of Aristotle has historically had a very great impact—but it remains, all the same, an error.

Now, consider another citation from Aristotle that is found in Book 5 of the same work: "A simple proposition is a statement, with meaning, as to the presence of something in a subject or its absence, in the present, past, or future, according to the division of time."

This text presents no difficulties, if we ignore the second half of the sentence; the simple propositions are the elementary propositions we have previously discussed, and the "somethings" that are present or absent are predicators. Only the additional "in the present, past, or future, according to the division of time" gives cause for thought. In Aristotle's sense, simple propositions include not only our elementary propositions of the form "x is p" but also propositions of the forms "x was p" and "x will be p." It is also possible to keep a single copula, "is," which expresses no tense and instead deal with tense by using a temporal indicator added to the proper name. However one decides to handle this particular issue, the critical appropriation of a traditional text must always be conducted using the appropriate methods of thinking.

In the case of a longer text, this critical appropriation can only be carried out in several operations:

1. Establishing one's own conceptual system
2. Critical reading of the text with consequent changes to one's own conceptual system
3. Subsequent reading of the text and possible further changes to one's conceptual system
4. Further cycles, as required

A person must thus repeatedly go from his own conceptual system to the text and then back again to his own conceptual system. Because a person's own conceptual system changes every time, however, this is not a circle but, rather, a spiral. Making a simple modification to Dilthey's phrase "hermeneutic circle," I would rather speak of a hermeneutic spiral. Naturally, this spiral is finite, as is every human endeavor.

This is as far as I want to go in my sketch of the most important steps of theoretical linguistic usage, of conceptual thinking. I hope that this sample will suffice for a concluding discussion of the following question: Does such a methodology of conceptual thinking offer the possibility of bringing an end to the multiplicity of opinion in theoretical questions?

To put the question in historical terms: What are the chances of a new Enlightenment through a logical-hermeneutical disciplining of thought?

Skeptics will answer: None at all, because, if there is now a chance of an Enlightenment, why has it not yet happened? Up to now, every attempt has run aground.

The fact of previous failures is undeniable. Otherwise, for example, we would indeed no longer have the apparently endless fights over principles among positivists, Christians, and Marxists. From this fact, however, we can conclude nothing concerning the future, if we can give reasons for the previous failures—reasons that existed in the past but now no longer obtain. Looking for such reasons, we must briefly review the whole of European intellectual history. The first enlightenment, the awakening of theoretical reason in general, occurred in the classical age of Greek philosophy. This first enlightenment did not succeed, because there were doctrinal disputes between Plato and Aristotle, disputes that today can be resolved but that were too difficult for autonomous reason as it was then taking its first steps. Greek theoretical philosophy came to an end in the skepticism of Hellenism and lost out in late antiquity and the Middle Ages to a dogmatic theology. Only in the Renaissance did reason begin once again to meditate upon its autonomy. This meditation, beginning in the seventeenth century, led to the modern model of freely created agreement, that is, natural science. Here we have a free monodoxy, obtained, to be sure, by restricting its investigations to that which can be disinterestedly ascertained. In contrast, the eighteenth-century efforts in the realm of the ethicopolitical have remained without success, because the methods employed in thinking continued to run in the tracks of the natural sciences. This applies as well to the nineteenth-century attempt at enlightenment, that is, Marxism. The popular conception of philosophy is an endless fight over theoretical questions of principle; unfortunately, this also was and is the conception of many professors of philosophy. It must appear very strange to this popular misunderstanding that in our time for the first time there should be fulfilled the one condition the lack of which has wrecked every previous enlightenment. This condition is the availability of a systematically teachable logic and hermeneutics. This synthesis of enlightened and historical consciousness was first attempted by Hegel but could not be completed until the work of Frege and Dilthey, to name only two. Once we have this synthesis at our disposal, it validates itself in that a person cannot argue against it without using it. Exactly therein lies, in my opinion, the opportunity for a new enlightenment.

TEN

Constructivism and Hermeneutics

Practical argumentation is a form of practical activity in that such argumentation is part of preparation for action. Formulated noologically, this runs: *Practical* thinking is that which leads to the formation of resolve (the setting of goals or ends) and the choice of means.

Practical thinking is always connected with our immediate circumstances. Rational argumentation leads to *intelligent* resolutions to act and to *prudent* selections of means.

Scientific thinking is underpinned and developed by the methodical improvement of practical thinking. It should lead to knowledge, understanding, and comprehension.

1. KNOWLEDGE

Argumentation over the nature of a given situation leads to formation of *opinions* about the situation and, in the case of rational argumentation, leads to *knowledge of the situation.*

Argumentation over the effects that will result from certain actions and that will change a situation leads to *opinions* about the effects of actions and, in the case of rational argumentation, leads to *causal knowledge* (in the natural sciences and the "empirically" conducted cultural sciences). In this connection, mathematics and protophysics are a priori auxiliary sciences.

2. UNDERSTANDING

Argumentation over which goals are served by certain actions leads (1) to the *interpretation* of actions as means to ends and (2) to the *interpretation* of ends as means to further ends (the structuring of ends by using the primary/secondary relation). This kind of rational argumentation leads to the *understanding* of actions and ends (in hermeneutically conducted cultural sciences).

3. COMPREHENSION

Argumentation over whether ends are forbidden or allowed leads to *judgments* about ends and, in the case of rational argumentation, leads in the cultural sciences to *comprehension* of ends as the application of ethics (see below).

Scientific thinking is theoretical thinking. *Theory,* however, also includes (under such terms as "theory of science" or "philosophy") the theory of scientific (i.e., knowing, understanding, and comprehending) thinking. Philosophy (in this sense) provides explanation and justification of the norms of rational argumentation, particularly the norms of rational speech in general (*logic* as rational grammar) *and* the norms of argumentation over opinions, designations, and judgments insofar as these can in general be given explanation and justification (for all situations, especially "culturally invariant" situations). For these areas of *theory of science* I suggest the terms "epistemics" (theory of knowledge), "scopics" (theory of understanding), and "ethics" (theory of comprehension). Taken together, logic and theory of science constitute philosophy.

Logic and theory of science are based upon the context of dialogue within an *esoteric* group; within this context all linguistic means can be explained and justified at any time. The languages of such esoteric groups are called *ortholects.*

For groups that depend on *exoteric* listening and speaking and reading and writing in a natural language, logic must be supplemented by hermeneutics and rhetoric.

Arguments over how one should take something heard or read exoterically lead to *interpretation* of speech and writing and, in the case of rational argument, lead to a *basis for interpretation*. Hermeneutics explains and justifies the norms of argumentation over interpretations.

Put simply, hermeneutics investigates how we *should* read a text, when we read it "with scientific intent." "With scientific intent" means here: with the intention of learning something for one's own thinking, for one's own conceptual system—something that it would pay to know for one's future actions.

Not everyone is obliged to read texts with scientific intent. A person can also read a text for entertainment or out of pure curiosity. However, when we read texts in order to become clear about our own goals, perhaps in order to change previously pursued goals or to avoid being bound by unexamined goals, then we are reading with scientific intent.

1. A person who reads with scientific intent already knows something in a systematic manner. Such a person must have systematically mastered at least part of some science. The systematic mastery of some part of a science implies, strictly speaking, a comprehensive grasp of the relevant scientific language. Insofar as they are in a logically unobjectionable state, the languages of science (including philosophy) may be called *ortholects*. If we formulate the task of a "textual interpretation" as, say, "to translate into our own language" the text's units of meaning (e.g., morphemes, words, or sentences), then this obviously should be understood as if the task were to translate the text into our own ortholect. Wherever this is successful, the author of the text can be said to have *already* known what we know. Wherever this is unsuccessful, it can be said that the author of the text *still does not* know what we know.

This method of interpretation may be called *dogmatic*. It does not call our own ortholect into question. Despite all the lip service to the contrary, a dogmatic interpretation does not serve to improve our own knowledge.

2. The usual method for avoiding dogmatic interpretation is to reject the use of one's own ortholect. Under this method we are content with our received language and translate the text into this language, which has not been subjected to critical reflection. We then understand the text as well as we understand our own language. Because in this case our own speech is not criticized, this method of interpretation may be called *naive*. In contrast to a dogmatic interpretation, a naive interpretation has the advantage that we can learn something from the text. However, it has the disadvantage that, despite all lip service to the contrary, we can never acquire *systematic knowledge* in this way.

We may try to defend the naive method by saying that the author of the text himself spoke no ortholect and that thus only with the naive method can we avoid alienating ourselves from the author's "linguistic perspective." I accept both of these assertions. However, from this it follows only that the naive method is the appropriate method, if our goal is to approach as closely as possible the author's linguistic perspective. Here, however, we have assumed that the text is to be read with systematic intent. And I asserted only that naive interpretation is inexpedient *for that purpose*.

3. If we accept the foregoing critique of the dogmatic and naive methods of interpretation, there results a third method, which avoids the criti-

cal failings of the other methods. This third method is called *critical interpretation*—more precisely, the systematic-critical method of interpretation.

Because the reading of texts with scientific intent presupposes a systematic mastery of some piece of science, I assume here that the reader with scientific intent possesses a sufficient knowledge of logic (in the wide sense of a rational grammar) to know how ortholects (whatever they may speak of) are constructed. This knowledge of logic includes, expressed in key words: the exemplary specification of predicator terms and relational predicators; terminological rules and definitions; theory of logical operators; description theory; and abstraction theory. Logic is not a sacrosanct doctrine; its components are always open to critical evaluation. However, anyone who wants to read a text with scientific intent must read it using his current and best knowledge of logic.

In a critical interpretation we do not employ our own particular ortholect that deals with the subject supposedly discussed by the text. Rather, we suppose that the author of the text had an ortholect of his own and that he wanted to say something in this *author's ortholect*. This supposition may in fact be false. Many authors have apparently had not the slightest thought about an ortholect. Other authors, especially philosophers like Plato, Aristotle, Hobbes, Leibniz, Kant, and Hegel, have wanted to a greater or lesser degree to replace their received language with a better instrument (here called an ortholect).

A critical interpretation takes as its task the job of "reconstructing" the author's ortholect from the text. This is partially the reconstruction of a fiction, because there was in fact no ortholect. Regardless of whether we call it a construction or a reconstruction, however, it is a matter of ascertaining from the text: whether the words are (insofar as they are not proper names) predicator terms, relational predicators, or logical operators; which exemplary specifications (direct or indirect) the author used for his predicators; which terminological rules he used; and so on. The answers to these questions are *not a translation* into our own language. They are only preliminary steps—steps that may be called *logical reconstruction*. By using logical terms (like predicator, definition, etc.) the critical method of interpretation avoids becoming dogmatic, because the logical terms are used to give the *description* of the author's systematically relevant use of language. The logical terms do not presuppose our own particular ortholect used for the subjects (supposedly) dealt with in the text. Logic texts (rational grammar) are an exception—this special case is put to the side here.

By using a logical terminology to describe the author's systematically relevant use of language, the critical method avoids both becoming dogmatic and the naive belief that we already have a basis for interpretation in cases where, for example, we interpret an author in the following way: For this author "ontology supplies the explanation and justification of the truth of logic" (as has been said of Picht, for example). In this kind of naive interpretation the logical reconstruction of the ortholect underlying the text (i.e., the deep structure as opposed to the surface structure of the text) is confounded with the task of translation into our own language. After the logical reconstruction of the author's underlying ortholect, critical interpretation confronts the following situations. First, the exemplary and terminological specifications that have been ascertained can supply sufficient information to fix a word (or sentence) of the author as synonymous with certain expressions in our own ortholect (which we have constructed on the basis of our own efforts to deal with the relevant subject). This would be a case of *translatability* into our own ortholect. Second, comparison of the author's ortholect with our own can show that the former contains certain terms (i.e., conceptual distinctions) that have eluded our own ortholect until now. In this case it is not possible to translate the text into our own ortholect, but we can expand our ortholect on the basis of the text's distinctions. Third, our attempts to translate the text into our own ortholect can lead to contradictions. In such a case we must systematically reexamine our own thinking and the author's results. Either we change our current ortholect on the basis of this reexamination (and then we would have learned more than we would from the second situation) or we reject the author's results.

Aside from these three cases, in which the text is sufficiently precise to allow a systematic dialogue with the author, by far the most common case is that the logical reconstruction is too imprecise for a systematic discussion. It may be, for example, that the text offers too few examples and counterexamples, or it may be that the terminological specifications are too obscure to be recognizable in the surface structure, and so on.

Whether in such cases we should give up the struggle with the text or whether (drawing upon other aids to understanding such as analysis of the historical, socioeconomic, and cultural background) we should continue with the logical reconstruction can be decided only in specific cases. In any event, it remains a naiveté to assume that we have already interpreted the text such that a translation into our uncritical colloquial or scholarly language has been produced.

ELEVEN

Rational Grammar

Rational grammar is a part of linguistics and so also a part of the science of language. In the following I shall therefore speak a good deal about speaking. This simple phrase, "to speak about speaking," indicates an exceptional situation; you cannot travel to traveling, or hear hearing, et cetera. All activities that have an object—we travel to a place, we hear a sound, and so on—are distinct from these objects. Normally, this is also the case with speaking; we speak of travels, sounds, and so on. We speak *of* objects that are *distinct* from speaking.

The exception is that we can also speak *of* speaking. Only in mathematics do we find something similar; we can construct mathematical theories concerning mathematical theories. This activity has been given the name *metamathematics*.

Anyone who speaks about language, however, does not speak a metalanguage; everything, including the ability to take itself as an object, is already a capability of the language of even primitive cultures in which we find no science or mathematics. *No matter of what* I speak, I will in any case *speak;* I will use language.

We use language in our daily life, poets use language, and scientists use language.

In the West scientific discourse is always able to appeal to tradition. When bringing up or introducing a scientific problem, one refers to the fact that even long ago—or at least recently, according to certain illustrious colleagues—precisely this or that problem was proposed as a particularly interesting problem.

In any event this is the case for our problem of a rational grammar. It is true that our European *grammatical tradition,* which begins with the Sophists, pales indeed beside Pasini's rational Sanskrit grammar—but analogous to it we have in the seventeenth century (to which Pascal and Leibniz belong) the example of the *Grammaire raisonnée* of Port Royal and a variety of projects for rational universal artificial languages. Today, linguistics in the forms of generative grammar and transformational grammar is essentially informed by modern logic. In fact, the majority of analytic philosophers of science accept logic as *the* syntax of all artificial scientific languages, that is, as simply *the* rational syntax. I should like to take this last point as the starting point for my discussion, so that I may—in opposition to this last point—speak in favor of a critical linguistics that would have a better understanding of itself. In this regard I will talk, in particular, about a rational syntax of elementary sentences that is not contained in logical syntax, and I will also talk about the problem of a rational semantics that is adequate to that syntax. To enliven gray theory a bit, I will deal with the normative aspect of color words.

In my concluding remarks, using the example of the tetrahedron of conceptual theory (fig. 11.1), I will try to show how through scientific

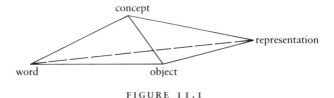

FIGURE 11.1

linguistics discourse could draw itself (to some extent by its own hair) out of the swamp of contentious discourse over concepts.

To this point my introductory remarks have appealed only to tradition. I have not adduced a problem but only some facts of the history of science that belong under the heading of "rational grammar."

As a scientist who specializes in theory of science, I should like to conduct an experiment here. I should like to try in a systematic manner to place the problem of "rational grammar" within the context of the sciences—and in a systematic manner to try then to place the sciences within the larger context of human activities.

To that end I will begin by considering *prescientific practical activities*. In such practical activities we find, for example, technical problems that can only be resolved with much reflection together with a systematic testing of experience. Thus all physical theories can be interpreted as supporting technical practices.

"Science," therefore, can be generally defined as theoretical support for practice. Because technology has only the task of producing means to the accomplishment of given ends, political practices, which first make possible agreement on ends, take priority over technical practices.

Political practices promote normatively ordered communal life. Man as "zoon politikon" distinguishes himself from social animals by the fact that for him there is no natural ordering of communal life. Man must give himself norms. In the simplest cases norms occur in the form of general conditional imperative declarations—so political order, without which men could not live as men, presupposes both logic and language. Now, whether or not it pleases us, we live in nations in which the changing of norms is constantly debated. We therefore have linguistic practices that are used in these debates over norms. These linguistic practices are a part of political practice—and therefore take priority over all technical practices.

The theoretical supports and underpinnings of these linguistic practices should, therefore, take priority over all technical theory. We might thus easily suppose that political science supplies theoretical support to the setting of norms (to "lawgiving," as it is usually called). Unfortunately, this is hardly ever the case; political scientists prefer to broaden conflicts. Not even the most politically engaged linguist would suppose that political science's desolate condition results only from the difficulties political scientists have in achieving a common understanding. It does, however, seem to me plausible to suspect that the linguistic misunderstandings between political scientists *aggravate* the difficulties inherent in achieving scientific consensus. And this suspicion immediately justifies considering how we might—for the purposes of scientific agreement—*reform* our received educated language (which is a more sophisticated form of so-called natural language). If necessary, we might even consider how we could replace this educated language with an artificial scientific language (perhaps something like scholastic Latin).

With this prospect in view I want now to turn to "rational grammar." We get along with each other relatively well in the West today; at any rate we live more or less well, although it cannot also be said that opportunity is justly distributed. For all those people who concern themselves with greater justice there remains enough to ponder and do.

On the international level, not even politicians, although masters of rhetoric, are able to talk away the injustices that exist.

The term "justice" presents no difficulties here. It does not need to be defined yet; it suffices to point exemplarily to situations that are labeled "unjust" by anyone who generally understands English. We know here what is referred to; such and similar injustices should become less frequent.

What now is to be done, if scientists who try to work out long-term guidelines for this task find themselves caught up in misunderstandings?

As scientists, they are obligated to argue scientifically. That is supposed to mean: For every definition and for every statement that they want to use they must argue *in a manner that is capable of producing consensus.* A consensus here means: universal approval, universal disapproval—or general forbearance from judgment. Put into the language of Roman jurists, consensus means: *sic, non,* or *non liquet.* If we ignore here the non liquet—in English it means "unclear"—then we are demanding that political science decide every relevant question. That, however, is not achieved by even the technical sciences. As regards the question of whether there are or are not infinitely many prime pairs, the answer of mathematicians is up to this moment: non liquet. Only sciences can afford this apparent luxury of the non liquet; in practice, be it technical or political, things must be decided. The requirement that a person produce arguments capable of producing consensus—that in cases of actual disagreement result in at most a non liquet, a general forbearance from judgment—distinguishes such argumentation from rhetoric that seeks to persuade others to accept an opinion. If the other does not agree, that is, if there exists a real disagreement, then in rhetoric that means only that each person sticks by his own opinion.

Actual disagreement among scientists means, however, that collectively for all participants the problem is not yet clear. The problem needs to be further argued—and argued to the point where no one any longer sticks by his own opinion. A person may privately have his suspicions or desires. It is, however, part of the discipline of scientific thought to acknowledge a non liquet in the cases of real disagreement. In this way each scientist recognizes and acknowledges the others as scientists. A person learns this disciplining of talk into scientific argumentation only to the extent that a person learns a science.

Let us assume then that two political scientists with the best intentions of speaking with each other in this way nonetheless get into some misunderstandings. As scientists, they want to tackle these hurdles together. What is to be done?

Without recourse to metaphor, what does this mean? It means that the two scientists should by linguistic means secure a common ground. Only after this has been done should they expand this common ground one step at a time until they can again speak about the problem that led to the difficulties. This procedure does not guarantee a resolution. It may lead only to a non liquet, but a non liquet that is a non liquet accepted by both to a question that both understand in the same way.

Assuming that these scientists are sensible enough at least to accept this suggestion, they will then begin to try by linguistic means to find a common ground.

Put into the usual educated language, they will try to find some common *categories*. That is what we call fundamental concepts with which further concepts are then definable.

If our scientists are unfortunately so educated that they immediately associate "category" with words such as "space," "time," "cause," "purpose," "being," or "becoming," then they will have no recourse but in the end to seek advice from professional philosophers.

I would recommend instead that they go to a historically educated linguist. He would inform them that the term "category" comes from Aristotle, who used it for syntactical distinctions only. Categories concern only the form of language, not the material (the content, as we say, using a distorted metaphor). We take therefore a common material; with that material we can next determine a common syntax insofar as all difficulties of content are avoided.

Because a *rational syntax* should be independent of the idiosyncracies of any natural language, we would like to imagine that, say, a Chinese scientist and a Russian scientist—I shall give them the names Mao and Leo—could try to make a fresh start and by syntactical means find a common ground.

As an unproblematic material let us take, say, a ball game. Everyone has played ball, at least as a child.

You can play ball without using words. In this prelinguistic activity we can "empractically"—as Buehler has called it—define the use of simple words.

Because I do not know which natural language the two scholars will use in playing ball, I will render their remarks using English roots. We could use, for example, German roots instead.

I trust you can easily imagine the sort of practical situations in which Leo would utter *imperative sentences* like the following:

(1) Throw!
Throw ball!
Mao! Throw ball!

or *indicative sentences* like:

(2) Ball does fall.
not: Mao does throw.

Because the material differences between, say, throwing and rolling are not of interest to us, let us pay attention only to the syntax. In 1 we made the acquaintance of imperative sentences, verbs and nouns, and also a proper name. These terms belong to a rational grammar, not just to the empirical grammar of natural languages. It is irrelevant, for example, that verbs are inflected differently than nouns in Indo-European languages. The noun's position *following* the verb is naturally also entirely a matter of convention. The distinction between the action and the thing (qua object of the action, as we say), however, is rational; the distinction is appropriate to the practice—in this case, the throwing of balls. Verbs are those words that are able to form imperative sentences by themselves; nouns complete such imperative sentences.

In 2 we made the acquaintance of indicative sentences, for which—by convention—a copula, "does," is used. Indicative sentences serve the function of orientation prior to further activity. Such sentences appear as either affirmative or negative, for which—again, by convention—no word is used for affirmation and a negative, "not," is used for negation. Internationally, in formal logic a special minus sign, ¬, is used as the written symbol for the negative.

That the ball as well as the man Mao appear as actors (which is obvious in English) is difficult to recommend as transculturally "rational." It is appropriate to direct the imperative "Throw!" at a man. It is not appropriate to direct imperatives at a ball. The ball does nothing but, rather, is simply in motion. Therefore, I would like to suggest two copulae for use in a rational grammar: a π-copula for action, a μ-copula for movements. I want terminologically to gather actions and movement together as *events*.

By definition, then, *history* consists of actions and movements of things. Action, movement, and thing are—to this point only—the categories of our rational syntax.

The actors of actions are men and animals; the distinction lies in the fact that men are acting as speakers, whereas animals are not speaking actors. Following an ancient tradition, we call men and animals taken together *living beings*. In this connection, it should be observed that the distinction of things that merely move into living (which are then called plants) and nonliving things is not a syntactical distinction. "Living" is not a category. It is justifiable only in biology, which gathers plants together with animals and men into the "family of living beings," because according to the theory of evolution the former are related to the latter; that is, they have a common ancestor. The distinction, living/nonliving, is self-evident in our culture, but the languages of other cultures need not have it. The distinction does not belong to rational grammar.

The sentence forms that have been "empractically" justified to this point can be extended further in various ways before we introduce logical operators. Here I will restrict myself to affirmative indicative sentences. The sentence, "Leo does throw ball," consists of a proper name, the π-copula, a verb, and a noun.

Extensions of such sentences, which are formed in English by using positional prepositions, can be rationally reconstructed: The ball is thrown, for example, "*to* Mao" or "*in* (a) window." At least 216 such positional prepositions can easily be given rational justification. These, then, are constructions that really amount only to reconstruction.

In addition, nouns can be used not only as so-called direct objects—as was done earlier—but also as indirect objects. This is done by producing justifications for grammatical cases, for example, the instrumental case (when the ball is thrown, say, "*with* [a] catapult") or the dative case (when the ball is given, say, "to Leo"). In any event, the acts of giving and taking already presuppose a cultural level in which property and exchange exist.

Which grammatical cases are available in a natural language and whether the cases available have their own case-morphemes, prefixes, or suffixes are matters of empirical grammar. A rational grammar can only reconstruct cases on the basis of whatever "practical employment" they have. In this respect a rational grammar always remains open to further extension.

The use of modifiers with verbs (understood to include actions and events) and nouns can be justified, namely, when it is a matter of additional distinctions within activity. For example, the ball can be thrown straight or at an angle, and the ball can be light or heavy. Such distinctions lead to "modifiers" that appear in Latin as adverbs or adjectives: verb modifiers and noun modifiers. Often these additional distinctions by modifiers of the activity are so important that it seems desirable to articulate in individual indicative sentences that the ball *is* light or heavy and that the throw *is* straight or at an angle. In English we use a special copula, "is," which is a convention. The distinction of this kind of sentence from the *event sentences* considered previously can, however, be given a rational transcultural justification. Therefore I want to propose for a rational syntax an ε-copula, so that we have as further indicative sentences such *specifying sentences* as "ball ε heavy" and "throw ε straight." The words preceding the copula are called subjects, whereas the modifiers are called predicates. Snell proposed in his famous *Aufbau der Sprache* a copula other than that for nouns or verbs to be used for modifiers as predicates that are used as kind- or species-predicates. We can nonetheless make do with one copula. I want only to note here that predicates are one- or many-valued.

These specifying sentences that consist of a subject and a predicate can also be extended, particularly if we now add a rational reconstruction of the words known in English as definite and indefinite articles. Setting this possibility to the side for the moment, we have then a lexicon that contains only four kinds of words and linguistic operators. The English sentence "Leo threw a heavy ball straight in a window" contains these four kinds of words and operators.

As regards attempts to reduce syntax to logic, I want to emphasize that this sentence is logically elementary, for it contains in rational reconstruction no logical operators.

Logical operators make their appearance only after we already have elementary sentences at our disposal. Words are taken from a lexicon and then put together according to a syntax. The simplest logical operators then combine elementary sentences into complex sentences. Therefore, these logical operators are called *connectives*. *Negation*, which is placed before a sentence, represents a special case. In this case, a logically complex sentence results from the expansion of a *single* sentence.

Connectives always combine *two* sentences. But on what basis? Where is the "practical employment" for such connections? I will here explicate only the case of "conditional junctor," which rationally reconstructs what occurs in English as the conditional structure.

The simplest case is supplied by conditional imperatives—for example, "If you see Mrs. Meyer, please give her my regards."

Ignoring the pronoun (which would be rationally reconstructed here as representing a proper name), we have two elementary sentences, the rational structures of which are already clear. These two sentences, A and B, are bound together by:

$$\text{If } B, \text{ then } A \text{ (symbolically: } B \rightarrow A\text{)}$$

What we need to do in order to comply with this request can easily be learned empractically. All legal norms—for example, the archaic norm "Murderers are punished with death"—are general conditional imperatives. That is, there is presupposed in the example a legal organization in which the following imperative holds for a judge:

$$\text{"If } x \text{ is a murderer, punish } x \text{ with death"}$$

These imperatives are universal because they are formulated with a variable, x. Any proper name may be substituted here for x. More precisely, we have here a *form* for imperative sentences that becomes a sentence only by prefixing the universal quantifier, "for all x."

Quantifiers, the universal quantifier and the existential quantifier, are logical constants that turn sentence forms into sentences; the variable in the sentence form is bound by means of the quantifier.

That the variables in legal norms are variables taking proper names is logically irrelevant. It is true that the standard logic textbooks always introduce quantifiers only for subject variables in elementary specifying sentences in which a predicate is attributed to one or more subjects, but that is a dogmatic restriction on the syntax of elementary sentences, which remains dogmatic even when the dogma is called "logical atomism" or even an "ontology." If, for example, an office supply store announces that felt pens are "available in all colors," then we have a universal quantifier that binds a variable for color words (which are adjectives, thus modifiers of nouns): "For all x, x-colored felt pens are available."

Besides the two quantifiers and negation there are in addition to the conditional connectives two more connectives: *conjunction* and *disjunction*, for which "and" and "or" are usually used in English. If we have implication at our disposal, then conjunctions and disjunctions can easily be given justifications. For example, I have previously made use of disjunction. After actions and movement (which were only exemplarily introduced) I defined the term "event" by saying that "event" is supposed to stand for "action *or* movement." For the nonlogician it is sometimes confusing that we can also formulate this same definition in the following way: "Actions *and* movement are called events." In fact, the latter is a misleading formulation. We do not mean that everything that is an action and (at the same time) a movement is called an event. Rather, all actions are to be called events, *and* all movements are to be called events.

This conjunction of two conditional sentences

(1) If something is an action, then it is also an event
(2) If something is a movement, then it is also an event

is equivalent to the single conditional sentence

(3) If something is an action *or* a movement, then it is also an event

This equivalence, demonstrated by using the colloquial usage of "and" and "or," is at the same time the practical justification of the use of disjunction.

With the six logical operators presented (negation, three connectives, and two quantifiers) logic is complete; all further logical investigations conform to the intellectual-economic use of these six operators. It remains to sort out, however, whether the theory of equality (particularly identity) including the theory of designation should be officially included in logic—or should be taken as a theory whose generality only extends to wherever synonymity is used.

The latter question concerns a dispute that is only terminological. In any event, the logical operators, particularly the quantifiers, are applicable to elementary sentences that contain no proper names.

With the preceding observation I will leave the empractical justification of the logical operators with the hope that it has become clear that logic is something added onto the rational syntax of elementary sentences. Logic is not a substitute for rational syntax; rather, it is an extension of it.

Having dealt with rational syntax, we must now consider *rational semantics*. Rational semantics can be studied using the example of a store with felt pens in all colors. Suppose you were to ask for a reddish yellow felt pen. Let us assume you are given a pen, while the clerk mutters something about "orange." Next, you ask for a yellowish red felt pen. Very probably a discussion of semantic problems will ensue. The clerk will argue, perhaps, that the color is called "orange" and "reddish yellow" is only a highfalutin English synonym, just as is "yellowish red"—the latter term having been favored only by the likes of Goethe. Or perhaps you meant, say, "brown"? If you were to deny this and say that to you brown is a *dark* reddish yellow, then you would perhaps be shown a light brown pen—and you might then counter that in your opinion light brown is a mixture of reddish yellow that indeed contains white but also contains *black* and is in that sense too dark for you.

There is a circle that represents a rational semantics for color words. Schopenhauer once demanded that a book by a colleague be legally banned because in the book the colleague had ventured to offer "greenish red." Syntactically, the proposal was unobjectionable. Besides syntax, what authority regulates the use of words? Syntax is taught in all schools, as is spelling; why is there no officially sanctioned semantics?

As far as color words go, we are well off here in the West, because we clearly have four major colors as sources of light: red, yellow, green, and blue.

Since the time that physicists began to meddle with our educated language, it has been considered rational to arrange the pure colors (i.e., those without a proportion of white or black) on a circle so that the complementary colors (those that when mixed result in white) are opposite each other. This can easily be done with the four major colors in English

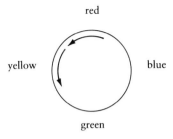

FIGURE 11.2

(fig. 11.2). (In Goethe's arrangement, red is at the top with colors of increasing frequency following counterclockwise.)

Schopenhauer, then, was rationally in the right with his proposed ban (whether the police should be available for its enforcement is another question). Physically, all transitions from red to yellow, to green, to blue, and back to red are continuous. How many differentiations would be reasonable? In his investigations into the customary perceptual effects of colors, Goethe called for differentiating the complementary interval of blue and yellow. With that we move from a circle of four to a circle of six (fig. 11.3). Yellow is now complementary to reddish blue (also called violet)—and that is justified by violet's being the last visible spectral color (before ultraviolet). With this circle green receives the status of an intermediate color, and thus we have Goethe's color triangle (fig. 11.4), an equilateral triangle having on each of its sides an intermediate color.

Goethe's supplementary justification for this triangle, namely, that blue is muddy black and yellow is muddy white, is unfortunately false. Here

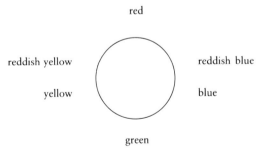

FIGURE 11.3

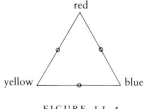

FIGURE 11.4

Goethe was mistaken, although in color theory he was otherwise a thousand times better than the physicists.

In any event, we can say that etymologically blue means darkly gleaming and yellow means lightly gleaming. Whereas "red" is a color word found all the way back into Indo-European and "green" is related to "grass" (and therefore is linguistically determined as grass-green), "yellow" is related to "gleam"; the Indo-European root—according to the Indo-European etymology in the *American Heritage Dictionary*—stands for "to shine; with derivatives referring to colors." Regarding "blue," the relevant Indo-European root also stands for "to shine." The Byelorussians are not blue Russians but, rather, white. There is no relying on etymology if we seek a rational linguistic usage. If we differentiate not only the pair blue–yellow (as did Goethe) but, in a second step (as has been customary since Helmholtz), also the pair green–red, then a circle of eight results. In figure 11.5 red is distinguished as the first visible spectral color (after infrared). The intermediate color blue/red is traditionally called "purple." Leaving purple to the side, there remain the seven colors of the rainbow.

The four major colors now form a color trapezoid with four intermediate colors. Set on its base, the trapezoid looks like figure 11.6.

To resolve the dispute between the proponents of the circle of six and those of the circle of eight, Ostwald suggested twenty-four divisions, because twenty-four is the smallest common multiple of six and eight. This

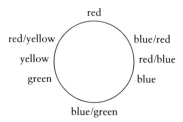

FIGURE 11.5

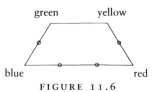
FIGURE 11.6

does not help at all with the semantics of color words, because, for example, it does not determine which of the colors between violet and blue/green should be called "blue." It seems to me that this is no longer a meaningful question. In any event the color trapezoid is an appropriate differentiation of the original circle of four. Had the physicists and their theory of the rainbow never come into this, then the circle of four would have sufficed in any case. In practice, only color words that belong to special things—for example, violets, purple snails, or turquoise stones—are added to the major colors. And, if one does not become infatuated—like the physicists—with pure colors and takes rather, for example, the adulterated colors that occur much more frequently, then a degree of intensity along the circle of the four major colors suffices in my experience.

As illustrative of the problems encountered in setting the norms of a rational semantics I have taken the case of words used for colors; similar problems arise in almost every area in which men speak. Indeed, although "semantic strategies" are the latest thing with politicians, these strategies were already successfully practiced in antiquity by the rhetoricians. Only for scientific politics does there arise the problem of a *rational* setting of norms for such words as "freedom" and "justice."

The example of color words shows, I hope, that a rational semantics is restricted to *definition*. In discussions one constantly hears one person first insist upon definitions and the other then respond with the true statement that one cannot define *all* words; some words allow, in fact require, undefined usage. This latter statement is true, but it is self-evidently true only because the term "definition" is used here in such a way that the definition of a word necessarily presupposes other words with which the word is defined. Empractical training in the usage of a word does not count as a definition. But "undefined" words are precisely a matter of such training.

Nonetheless, definitions are obviously expedient for a rational semantics; they contribute to the economy of speech. Scientific terminologies also make extensive use of definitions.

In natural languages definitions are rare. A clear case of definition is the designation of certain family relations:

brother-in-law = df. the brother of a spouse or the spouse of a sister or brother, e.g., if two sisters are married to two brothers

uncle = df. the brother of a mother or father or the spouse of a sister of a mother or father

The predicates occurring in the definiens—with the exception of a two-valued predicate, "spouse"—can be further reduced by definition to a two-valued predicate, "child" (x is the child of y), and two adjectives, "male" and "female"; for example:

x is the mother of y = df. y is the child of x, and x is female
x is the father of y = df. y is the child of x, and x is male

A brother of x is then a male child with the same father and the same mother as x (but distinct from x). "Sister" is defined in a similar manner.

The result of such definitions is a group of words that can be designated as basic words. These words are the words that remain after all definienda have been replaced by their definientia.

If scientists, such as Mao and Leo in our example, were to work back to a common practice (if necessary, to ball games) and then together were to empractically train themselves in the use of the basic words, using examples to move on through the setting of semantic norms and definition and thus beyond mere syntax, then we could also request that political theorists (practical political rhetoric I leave to the side) come to a consensus over the usage of words like "freedom" and "injustice." The consensus achieved could be, for example, a non liquet: Men might come to agree collectively that the word "freedom" is not to be used until further notice. In this way rational grammar would be a fundamental discipline that would be very useful, particularly for political science.

But today such a prospect is "utopian"; that is, it cannot be attained in either the short or the medium term.

As is well known, we should not predict anything over the long run, because we have only to set the end point far enough in the future and anything possibly possible becomes attainable.

At present there is a problem that has particularly thwarted attempts by linguistics to construct a rational grammar: This problem centers on the tetrahedron of conceptual theory I mentioned earlier (fig. 11.7). In conclusion, I want to try to delineate this problem precisely.

Up to now, I have spoken only about words and objects, understanding "objects" to mean occurrences and things.

We use words to speak about objects. If you stand before a tree and say

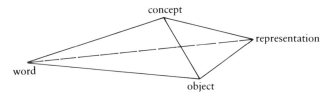

FIGURE 11.7

"tree," that act represents the first dimension, the dimension of unproblematic practice. Everything is, as we say, concrete.

But the reader of this presumably is not really standing before a tree. You only *imagined* a tree. This act is still relatively free of problems: an elementary piece of the psychology of imagination, a second dimension that is learned by everyone who learns a language. We learn to speak not only about that which is immediately present. For example, we remember the past, we have representations in memory—and about these we speak as if we were still in the past. Apparently, animals also have memories, but they do not speak about them. Therefore, they also forget them more quickly, and above all they do not share the joy and pain of human fantasies. We take part in the sharing of memories through words and sentences. Simply by combining words and sentences in novel ways we are able to produce new representations. This is how we fantasize—and we can also speak about these fantasies; we can share them.

Representational activities, whether memory or fantasy, are human activities that, like every other activity, happen at particular times. A man finds himself in a particular place and then imagines, for example, a tree; shortly thereafter he imagines something else or even nothing. That is what we mean when we say that representations occur in space and time. And therefore we call such representations *concrete*, although the representing itself is an *inner* act in contrast to an *outer* act, in which we move our bodies in space and time.

The triangle of concreta—word, object, and representation—conceals no serious difficulties. Things become difficult only when the word "concept" is added. We must then raise ourselves above the level of concreta and into a third dimension, into the dimension of abstract concepts. Whatever concepts may be, they are not concreta. Therefore, they are called *abstracta*. But the word "concept" is concrete, and the difficulty lies in determining and explaining an appropriate usage for this word.

The solution that is unfortunately almost universal is to turn concepts into concreta. Either one suddenly finds certain *words* called "concepts," particularly when it is a matter of the terms of a scientific language, or one finds that the *representations* belonging to these words are called "concepts." If the words are called "concepts," then the corresponding representations are usually called "conceptual contents." With all this, however, we obviously remain on the level of concreta. The movement up into the heights of abstraction has not yet occurred. These observations are not a solution to the problem; they only make the difficulty clearer.

In any event, each person is free to indulge himself in a philosophy in which he asserts that there are only concreta. We live in a pluralistic state of law. A person is legally entitled to hold whatever peculiar philosophy he likes.

Things were different in the Middle Ages. Anyone who ostentatiously consorted with nominalism, the doctrine that besides concreta, objects, or representations there were only words (nomina sunt flatus vocis), was suspected of heresy.

In those days you were certain when you asserted that concepts were a second kind of representation. In addition to the usual representations, called individual representations, there was a second kind, so-called general representations. Concepts were of the latter kind.

This sort of attempt to classify concepts psychologically has remained popular to the present. Today, it is called "psychologism." Since Frege and Husserl, psychologism has stood scientifically refuted, but, because we do not prosecute inquisitions into erroneous doctrines, this doctrine is still frequently encountered. Dictionaries, for example, give "representation" without any qualification as a meaning of concept.

Finally, there is also the claim that concepts are a second kind of object in addition to concrete objects. This doctrine also had its proponents in the Middle Ages. It was called realism after the Latin word, *res*, for object. Realism remains today the dominant doctrine among mathematicians—and therefore also, for example, among mathematical linguists. Mathematicians do not deal with objects; rather, they work only with symbols. They claim, however, that their symbols are the proper names for a kind of object accessible only to them, abstract objects like numbers, sets, and functions.

This realism is usually involved in the dispute over the Platonic doctrine of ideas, according to which geometry deals with objects of a special kind; as ideas, the surfaces with which geometry deals are distinct from all real surfaces—wood, steel, or glass, for example.

Aristotle opined that ideas were concepts—and Aristotle was a psychologist, or at least he was no realist. So it has come to pass that realism is also called Platonism.

You can see how, by trying to rise to the heights of abstraction, we instead become mired in a swamp of scholastic disputes. In my attempt to find a way out of this swamp I shall set geometry to the side and thus also the doctrine of ideas. We will restrict ourselves to concepts, and that means we will try to determine and explain for the word "concept" a usage that does not imply that concepts are concreta.

The explanation that I will now sketch is usually presented under the heading "abstraction theory." Precursors of this theory can be found in Aristotle and Leibniz and also, more recently, in Frege, but constructive theory of science was the first consistently to take this theory as a basis for all abstracta, for example, the sets used by mathematicians.

Here we are concerned with abstraction theory only to the extent that in educated speech we speak not only of the word "man" but also of the concept "man"—not only of the word "justice" but also of the concept "justice."

According to abstraction theory, this talk of concepts is to be reconstructed on the basis of the *synonymity* of words. But here we encounter a difficulty. Defining under what conditions two words of a natural language are synonymous appears a hopeless undertaking. It is easier with words from different languages. For example, the English word "Saturday" is translated into High German as *Sonnabend*. This kind of translatability, which presents no difficulties in the case of days of the week, is also called synonymity. In addition, "Saturday" is not etymologically related to "Samstag" (the day of the Sabbath). Both of the German words are synonymous with each other, because two words that are synonymous with a third are synonymous with each other.

The last statement in the paragraph above is a theorem that results from the definition of synonymity. Such a definition cannot be found for words of natural language; against every proposed definition there are always counterexamples.

For a rationally reconstructed scientific language that would contain a rational semantics as well as a rational syntax and explicit norms of word usage, synonymity can be very easily defined. Two words are synonymous if they have the same valid norms of usage.

For example, if you translate from one scientific language into another scientific language, you need only to check whether in the translation all semantic norms of the one language correspond to the semantic norms of the other language and vice versa.

Within a scientific language it is inexpedient, however, to use synonymous words. Ockham, the great nominalist of late scholasticism, recommended that one not unnecessarily multiply entities.

Let us take, for example, some of the terms of a rational syntax: "action," "event," or "thing." In translation into German we would have *Ding* instead of "thing." The word *Ding* has four letters, whereas the word "thing" is fortunately not a four-letter word. Intellectually, we can say that the number of letters concerns only the words and not the concept *conveyed* by both of two synonymous words.

This talk of concepts is easily reconstructed. We produce statements in which a term of a scientific language occurs—here, for example, the word "thing." Some of these statements—for example, those concerning the number of letters in the word—are true for the term in one language but not true for the synonymous term from the other language. Such statements are called *not synonymity-invariant*. With this point explained, we then restrict yourselves to synonymity-invariant statements about the terms of a scientific language.

For example, if the term "entity" is translated into German by *Gegenstand*, then "things are entities" is a synonymity-invariant statement. It corresponds to "Dinge sind Gegenstaende."

It is with the restriction to synonymity-invariant statements that we bring ourselves up into the heights of abstraction. We abstract from the phonetic form of the synonymous words, with only the usage norms being relevant. Now, in the case of a synonymity-invariant statement A about term T, instead of explicitly adding that this statement is synonymity-invariant, we reformulate the statement A into a statement about the concept T. Because "things are entities" is synonymity-invariant, we reformulate this statement as

> Thing is a subconcept of entity

We have now reached the heights of abstraction. The process of abstraction performed here, namely, the restriction to synonymity-invariant expressions, is not a psychical process and is not a process involving nonlinguistic objects. The process of abstraction is a linguistic process.

It has even been said that this is a linguistic trick. Quine, the famous American logician, speaks in this connection of "innocent abstraction," because we speak here of concepts only *as if* they were new objects in distinction from words and concrete objects. The ready response to this is that the mathematician deals with more than just abstract objects that result from a fiction. Against this response, constructive mathematics demonstrates that we do not have to agree either to psychologism or to realism in order to engage in modern mathematics (to which I must add: to the extent that modern mathematics is useful).

As regards the problem of a rational grammar, the only important point of this foundational dispute among mathematicians is that we not be irritated by the dispute. In a rational language, talk of concepts is easily introduced *per abstractionem*. That should satisfy the linguists.

In making the reflective move to a rational grammar, that is, to rational syntax and semantics, the linguist brings himself into a region that lies between the regions of empirical linguistic research and mathematics.

In this region a person runs the constant risk of dogmatically using the methods of one of these bordering regions. It seems to me, however, that despite this danger rational grammar should be studied, because in the context of science rational grammar is a foundational discipline for all political science that involves rational argumentation over political norms.

PART IV

Philosophy of Mathematics: Arithmetic

TWELVE

How Is Philosophy of Mathematics Possible?

My title contains the words "philosophy" and "mathematics." I need first to explain the meaning I attach to these words in this investigation. The degree of difficulty of our problem here is dependent upon the meaning that we give to these words. We could make the problem easier, for example, if from the outset we were to view mathematics from a fully determined aspect—if we were to admit as valid mathematics, say, only certain axiomatic theories, as is done in the modern French encyclopedia of mathematics that is published under the pseudonym "Bourbaki." Of course, I would find it even more congenial if we were to count only operative mathematics as true mathematics. But I will reject this course; I would much rather simply circumscribe mathematics as a historical phenomenon, so that our topic does not unnecessarily lack content. Because our question is directed at the present, our investigation is restricted to contemporary mathematics, let us say the last hundred years of mathematics. The last hundred years in mathematics represent a complex of scientific activity that more or less hangs together. No matter how vague its outer boundaries may be, this complex has arithmetic and analysis, that is, the theory of numbers and their functions, as its distinct nucleus. Deposited upon this nucleus are axiomatic theories that draw their models from the nucleus. Examples of such theories are, above all, algebra and topology—but even theoretical geometry can be subsumed here and can thereby assert membership in mathematics against all the voices that want to drag geometry down to a physical theory.

What makes philosophy of mathematics possible? The question depends also upon the meaning that we give to the word "philosophy." Here we cannot use the method of historical circumscription, as we did above. That which has appeared under the title of "philosophy" during, say, the last hundred years does not, even when viewed sympathetically, reveal a complex unity, and we can discover a nucleus only by importing our own position. We would need to choose the genuine philosophy from the accumulation that is the entire history of philosophy, because a restriction to the last hundred years would be totally arbitrary here, and this necessity offers an opportunity to push our question in the direction of the trivial. For example, since antiquity logic has been part of philosophy, even if it has appeared under different names—dialectic in Plato, analytic in Aristotle, and *Vernunftlehre* in German academic philosophy. What could be more natural than to understand philosophy of mathematics as the theory of the logical principles of mathematics? There would be no question that such a philosophy of mathematics is possible. It is just this obvious possibility that indicates that this solution would not be a good answer. This "philosophy" would be a specific individual science, because no one can deny today that logic has achieved the independence of a science. Philosophy as the mother of all sciences has already seen many of its daughters to the door. Logic has the advantage of being allowed to continue to live in the stimulating atmosphere of the maternal house. Its independence within the house is, however, nothing new. We might say that from its birth, namely, Aristotle, it has occupied a special position in that logical thinking is a presupposition of every science. Because all philosophical thinking must follow the rules of logic, logic is also a precondition of philosophy. Consequently, Aristotle's work comes to us with logic posited as the tool, as the organon that precedes everything else.

So, if you want to speak of philosophy of mathematics, you are justified from the modern perspective in simply equating logic, insofar as it is used in mathematics, with mathematics. Viewed from the perspective of the modern sciences, the problem of a philosophy of mathematics appears more or less as follows. Mathematics (including logic) goes confidently on its way; it progresses. But occasionally this progress hits upon certain difficulties for which the usual methods are no longer useful. Indeed, these are the situations in which difficulties in fact resist precise formulation. One hits upon difficulties involving "principles"—and such difficulties must certainly be philosophical ones. So philosophy is given the honor of being allowed to serve, so to speak, as the curio cabinet of the individual sciences. This is probably well intentioned; it is probably feared that philosophy would be otherwise unemployed, if it were not kept occupied in this way.

Nonetheless, I would like politely to request that the individual sciences look after their own curios, no matter how principled these curios may be. For example, when mathematicians want to know about so-called natural numbers—for example, to what extent they are "natural" or whether the customary axioms for natural numbers are entirely valid—how should a philosopher know these things, if the mathematicians, who work constantly with these numbers, do not know them?

Good. If we will not allow boundary and foundational questions, any more than logic, to be the substance of philosophy of mathematics, then our question concerning the possibility of mathematics becomes even more pressing. Not how such philosophy is possible but, rather, whether such a thing is possible, whether such a thing can be at all—that is the question to be asked now. Up to this point we have only approached the issue negatively and determined that

1. Logic should not be called philosophy of mathematics.
2. Philosophy should not deal with the questions that arise in the individual sciences.

If we do not want to be completely arbitrary and call "philosophy" just any old thing that we might pull out of the unsurveyable multitude of remaining possibilities, then only the method of historical determination is left. How, then, has it come about that a phrase like "philosophy of mathematics" has not been viewed as absurd but, rather, has given some kind of intelligible, even if very unclear, direction to our thinking?

Here let us return to Aristotle! According to tradition, the *Prior Analytics* and the *Posterior Analytics* are followed by the works concerning the individual sciences, collected together as *Physics*. After that begins the "royal road of philosophy" from πρώτη φιλοσοφία to *Ethics* and *Politics*. πρώτη φιλοσοφία, first philosophy—so named because of its position, μετά τα φυσικά, after physics, thus "metaphysics"—inquires into first causes and reasons, into the principles of all those things that exist. Our world, however, is no longer the world of the Greeks. Therefore, we can no longer identify philosophy with metaphysics in the Aristotelian sense. We do not need to analyze here the events of intellectual history that destroyed the Greek unity of thinking and being and that finally, following the transit through medieval theology, led to our understanding of the world that has been formed by empirical natural science. Kant was the first to see that metaphysics in Aristotle's sense had become impossible. In place of the metaphysical question concerning the principles of existing things, Kant brought onstage the question concerning the principles of our knowledge of existing things. Transcendental logic took the place of metaphysics.

In this way, our historical determination of the meaning of "philosophy" leads us to Kantian philosophy of mathematics, that is, the transcendental question, What makes pure mathematics possible? Because we want to restrict ourselves here to the nucleus of contemporary mathematics, arithmetic and analysis, we need look only at Kantian philosophy of arithmetic. As regards analysis, the foundation of which was even more incredibly confused in Kant's time than it is today, Kant appears to have adhered to our previously formulated maxim—not to view philosophy as a curio cabinet. He left it to mathematicians to struggle with analysis. As regards arithmetic, he destroyed the Hellenistic doctrine, which had been accepted by Leibniz, that numbers insofar as they are ideas are the thoughts of God. Since Kant, God has ceased to be a mathematician. Mathematics becomes as it was in late medieval nominalism—much more an instrument of finite man. It seems clear to me that this conception has become widely established today. It has become self-evident. In contrast to Gauss, who expressed the (at least verbally) unclassical proposition "ὁ θεὸς ἀριθμητίζει," Husserl remarked, "I would say at most ὁ ἄνθρωπος ἀριθμητίζει." Now, this does not at all mean that today Kant's transcendental-logical discussions of arithmetic are a recognized part of the philosophy of mathematics we seek. On the contrary, the Kantian doctrine that arithmetic propositions are synthetic a priori finds defenders only among a few true Kantians. The tremendous cumulative attack made by Frege and Russell, the goal of which was to establish using the additional methods of modern logic that arithmetic is analytic, has survived the crossfire from Brouwer's intuitionism and Hilbert's formalism. But neither Brouwer nor Hilbert was a strong Kantian. The boundaries between logic and mathematics have been so blurred by these discussions that it appears to be no longer meaningful to mark off within the region of the logico-mathematical a subregion of that which is analytic. Moreover, the Kantian doctrine of the relation of numbers to time has become completely meaningless for foundational research in mathematics. Even for Kant this doctrine acquired its significance only when it was placed within the complete corpus of transcendental logic. This happens not when we ask the question "What makes mathematics possible?" but, rather, when we ask the question "What makes a priori knowledge in the natural sciences possible?" This question lies outside our present topic. Therefore, I will give only a brief report of the historical fate of Kantian philosophy. Kant separated the formal sciences, logic and mathematics, from the individual empirical sciences, the factual sciences. Kant's transcendental logic provided the connection between them. Even in Kant's lifetime this transcendental logic began to be misused in German Idealism as a reincarnation of the old metaphysics on the one hand and to be distorted into a psychological theory of knowledge on the other hand.

It is in these forms that transcendental logic has had its effects to date. Modern positivism even asserts that there is no gap between the formal and the factual sciences. Frequently, in grotesque ignorance of the historical intellectual events that have led up to the phenomena of the modern natural sciences (which resulted precisely from a unique fusion of mathematics and experiment), it is dogmatically asserted that every proposition is either mathematical or empirical. Therefore, something like a synthetic a priori cannot exist. At this point we may leave open the question of how we could remove this mountain of misunderstanding that blocks philosophy from being conceived as the theory of synthetic a priori propositions. For our purposes here, it is only important to note the fact that, due to the effectiveness of positivistic thought, the word "philosophy" has signified something pseudoscientific for the last hundred years, something like a mixture of science, poetry, and religion. Husserl's attempt to make philosophy a rigorous science did not succeed. So the only legitimate object that has been conceded to scientific philosophy is its own history. The nineteenth century, the century of historical consciousness, also produced scientific history of philosophy. That appears to be the end of all philosophy.

Things appear to stand rather poorly with the object of our inquiry: the possibility of a philosophy of mathematics. Our historical determination brought us to the post-Kantian reduction of philosophy to its own history. At the same time, however, precisely this historicization contains a new possibility for philosophizing. It was only by trying to understand historically the philosophical systems from Plato to Hegel that philosophy began (using the term "world view," since become obsolete) to reflect upon the ideas that men make of the world and of themselves and that are determined by all human activities, in particular the activity of thinking. But, as early as Fichte's pronouncement that "the kind of philosophy a man chooses depends on the kind of man he is," it was clear that man's possibility of making a picture of his existence-in-the-world was not simply an object for historical observation. Dilthey spoke of the "inability to go behind" life. No amount of detailed knowledge of history obviates the fact that we can only conceive and understand our own existence-in-the-world in precisely one way. We cannot live without some prior understanding of man and world. This understanding does not formulate itself in propositions that could be part of a science. To the extent that this understanding is not only expressed in practical decisions, it can only be grasped as a complex of conceptions that precede the sciences and that first arrange these sciences into a living whole. I want to call these conceptions more precisely *preconceptions*. Following upon the apparent demise of philosophy we can newly appoint to philosophy the task of reflecting upon these preconceptions. Philosophy has no preconceptions to represent; rather, it has preconceptions as its objects. This formulation is suggestive of Dilthey, of course, but it should not be taken to indicate that philosophy has relativistically to view all preconceptions as equally justified.

The use of the term "preconception" should guard against our taking as objects those conceptions that, instead of preceding science, come after science as ostensible philosophical consequences. Following the so-called collapse of Hegelian philosophy, Wundt proposed that philosophy be conceived as the synthesis of scientific results within a world view. In this way, there arose a metaphysics of "postconceptions" or afterthoughts. Since then, people have tried adaptively to fit these afterthoughts to the contemporaneous state of the sciences. This activity has so little in common with the earnest labors over the foundations of knowledge that from Plato to Hegel appeared under the heading of "philosophy" that I can see only a sorry caricature of philosophy in this metaphysics.

With this (I hope sufficiently clear) proposal to take the critique of preconceptions as the genuine nucleus of the history of philosophy, our question concerning the possibility of philosophy of mathematics becomes the more precise question concerning the preconceptions that underlie contemporary mathematics. But greater precision does not answer the question. It is easy to imagine mathematicians indignantly rejecting such a point of inquiry. What could uncontrolled and shifting prescientific conceptions have to do with the objective and demonstrable theorems of mathematics? Am I perhaps endorsing the sort of situation, which fortunately existed in German mathematics for only twelve years, in which one seriously proposes the possibility of a Western mathematics or something similar? That would be a complete misunderstanding. In the following, when I try to show the preconceptual limitations of contemporary mathematics, it should be understood that my intention is to transcend these limitations. Philosophy of mathematics is given the task of bringing to light the preliminary decisions that lie hidden in the initial steps of the construction of mathematical theories and doing this by placing them in the larger context of the history of prescientific conceptions. Only in this way can we bring these preliminary decisions to consciousness; only in this way can we find the motivation, the will, to carry through necessary changes.

Nonetheless, I do not want to recommend a program here. Rather, I

How Is Philosophy of Mathematics Possible? | 163

want only to discuss a few concrete instances in which the philosophical presuppositions of mathematics become clear. I will restrict myself to three cases:

1. the use of *tertium non datur* in arithmetic
2. the use of the concept of the power set
3. the use of axiom systems that have not been demonstrated to be free of contradictions

The issue here is a proposition, a concept, and a method that are in daily use wherever there are mathematicians. This use is a simple fact. We must see whether this fact can be made intelligible.

Anyone who is unacquainted with these things will certainly be inclined to think that, if mathematicians use this proposition, et cetera, then they must surely be correct in doing so. Indeed, we live in an age of scientific faith. But what if this use is nonetheless not completely correct? We will want to inspect these things more closely. First, the use of *tertium non datur* in arithmetic. Arithmetic is the theory of numbers. The cardinal numbers—1, 2, 3, . . .—and counting with them form the basis of all mathematics. According to Kronecker, these numbers were "made by the lord God." This benign expression is very often quoted in textbook introductions. But why? The intention is to move the reader not to reflect upon the origin of the cardinal numbers. The reader is supposed to accept these numbers as he does the sun, the moon, and the stars. They are simply there. Fermat claimed that, if n is some cardinal number greater than 2, then the equation $x^n + y^n + z^n$ cannot be satisfied by any three cardinal numbers x, y, and z. This can be expressed in a formula:

$$\bigwedge_{x, y, z} x^n + y^n \neq z^n$$

There are infinitely many numbers n for which Fermat's assertion has not yet been proven by anyone. All the same, no one can refute it either; that is, no one knows of a triplet that satisfies the equation. But, if the lord God made the numbers as he made the stars, then he would know whether there is such a triplet. In contrast, we know only that[1]

$$\bigwedge_{x, y, z} x^n + y^n \neq z^n \vee \bigvee_{x, y, z} x^n + y^n = z^n$$

It appears as if we could prove this formula without having to trouble the lord God. Somewhat simplified, this formula has the form

$$(1)\ \bigwedge_x \neg a(x) \vee \bigvee_x a(x)$$

---

1. $\bigwedge_x$ = for all x; $\wedge$ = and; $\neg$ = not; $\rightarrow$ = if . . . then; $\bigvee_x$ = for some x; $\vee$ = or.

In modern mathematics all formulas of this form are recognized to be logically valid. Therefore, the deduction

(2) $\neg \bigwedge_x \neg a(x) \to \bigvee_x a(x)$

is always admissible as a logical inference. This demonstrates how the inferrer understands both himself and his thinking. Namely, the totality of numbers is conceived here to be like a basket of apples. If you know that not all the apples in the basket are green, then you can obviously conclude that at least one apple in the basket is ripe. You need only check each apple in turn for ripeness. Should you find no ripe apple, then it would have turned out that all the apples were green—in contradiction of the presupposition. If you carry this way of inferring over to numbers, then you thereby forget that there are infinitely many numbers. You can check all of the apples in a basket for a property because there are only finitely many apples. No one can check all the numbers this way. Anyone who wants to check them will never reach the end. The number series has no end; precisely this is what constitutes its infinitude. Aristotle expressed this in the third book of his *Physics* in this way: "In general, the infinite exists only in the sense that we always take another and then again another. That thing which is taken, however, is always finite, and in each case a different thing." Here the infinite is only δυνάμει ("in potentia"), thinkable, not ἐνέργεια ("in actu"). It was incorrect, of course, for me to say previously that modern mathematics had forgotten that there are infinitely many numbers. Mathematicians know this very well, but they understand human thought as being capable of dealing with the infinite largely as it deals with the finite. Here we have a case where systematic discussion is unable to find a starting point; only the historical perspective is able to throw light on the situation. In antiquity, at least in post-Aristotelian antiquity, the actual was restricted to the finite. Only in the Middle Ages was the Judeo-Hellenistic representation of the infinite God combined with the pure actuality of God taught by Aristotle. It was only from that point on that we became accustomed to talking of an actual infinity. The Renaissance transported this actual infinity from the other world to this world. Modern physics, however, has arrived again at finite world models. In contrast, mathematics first saw actual infinities flower in this century with set theory, despite the beginnings made in the infinitesimal calculus in the seventeenth century. I must admit that an actual infinity seems to me an anachronism after Kant. In particular, it is only with difficulty that it can be reconciled with the otherwise accepted and fundamental instrumentality of conception in all scientific thinking. Historically, it can be explained only by the philosophical indifference

How Is Philosophy of Mathematics Possible? | 165

into which mathematics, along with natural science, has fallen. Thus the effects of Kant have not been widespread. For example, the philosophical background of Russell is much more suggestive of Leibniz than of Kant. Even scholastic influences can be seen in Bolzano and Cantor. I do not want here to discuss systematically the validity of formula 1, above. That should be left to mathematics, as should also its deduction from the logical principle "A or not-A, tertium non datur." The latter is expressed in the following formula:

$$A \vee \neg A$$

Brouwer, who in 1907 first recognized this connection, did not hesitate to discard tertium non datur, $A \vee \neg A$, for arithmetic. That he was not followed in this is explained not by the often emphasized complications that result from the abandonment of tertium non datur but only by the traditional conception of human thinking as a recapitulation of divine thinking. All of modern mathematics rests upon this conception—here it is completely clear that mathematics depends upon a preconception. If you free yourself from this dependence—and to do this it is essential that the origin and history of the dependence be brought to consciousness—then you will find within the customary so-called classical logic a nucleus that is independent of the "finite/infinite" distinction. It happens that this nucleus is sufficient to prove that the employment of classical logic is free of contradiction. To that extent we can say that a critique of tertium non datur does not decisively affect modern mathematics.

Our second case constitutes an essentially more critical matter, the use of the power set. Here the issue is as follows. Let us again begin with cardinal numbers. These represent a totality, a set, to use the special term. The squares of the cardinal numbers also form a set, as do the even numbers, the prime numbers, the numbers that are greater than 5, and so on. These sets are defined by properties. For example, the set of prime numbers is defined using the property *prime*. X is prime, if the following is valid

$$(3) \; \bigwedge\nolimits_{m,n} \times x = mn \rightarrow m = 1 \vee n = 1.$$

The transition from a formula such as 3 of the form $A(x)$ to the set of x for which $A(x)$ is valid is unproblematic. This transition is necessary only because various formulas can represent the same set; for example, instead of 3 we could use the following formula:

$$\neg \bigvee\nolimits_{m,n} \times m > 1 \wedge n > 1 \wedge x = mn.$$

Sets are produced by abstracting from the multiplicity of formulas. All of the sets considered contain numbers as elements, so that they are contained as parts in the set of all cardinal numbers. Obviously, there are infinitely many such partial sets. Therefore, modern mathematics also speaks of the totality or set of *all* partial sets. That is the power set. We do not want to concern ourselves with the propositions that have been arrived at concerning this power set. Rather, we want only to examine critically the construction of the concept "power set." A systematic definition of this concept is not available; only some examples of partial sets are given with no prescription for obtaining all partial sets. Nevertheless, the concept is in use.

One of the most frequently read textbooks of higher mathematics, Mangoldt–Knopp, says in its newest edition of 1955: "It is not possible to explain in the true sense the concept of a set, i.e., to reduce it to still simpler concepts. Our intellectual ability to mentally gather together certain things into a unity, a set of these things . . . , must rather be seen as one of the fundamental capacities of our intellect." To be sure, here sets are not—as was the case with numbers—made by the lord God; rather, it is the intellect that makes them. Practically, however, there is no difference, as it is supposed to be a matter of an inconceivable creation from nothing. A study of the (to a certain extent naive) mathematical literature, of which Mangoldt–Knopp serves as a representative example, immediately reveals that two things have been thrown together. Under critical examination the abstraction that allows the transition from a formula $A(x)$ to the set of x for which $A(x)$ is valid—symbolically: $\varepsilon_x A(x)$—is easily seen to be an unobjectionable *façon de parler*. But the fundamental capacity of the intellect to form sets is appealed to not only in the justification of this abstraction but also in forming a set from *all* partial sets. The latter is something else entirely, because it amounts to one's claiming to produce *all* formulas $A(x)$. The use of the concept of the power set signifies the use of the concept of an "arbitrary formula." For this, the alleged fundamental capacity of abstraction is of no use. What one is thinking of when one speaks of an arbitrary formula $A(x)$ may perhaps be this: A formula $A(x)$ is a prescription, constructed using the linguistic tools of mathematics, in which the variable x appears. In these "linguistic tools" lies the heart of the difficulty. Mangoldt–Knopp says explicitly in one place that it is not necessary that this prescription be a mathematical formula; it is sufficient if it "is given only in words."

I beg your pardon for the digression. What has just been said concerning the power set is of course not philosophy of mathematics but only an elucidation of mathematical matters, which is necessary as preparation. The philosophy that inquires into the preconceptions that make the bases of the special sciences intelligible first begins when we ask how we are to explain that mathematics views the totality of its linguistic tools as given from the outset—in any event, as circumscribed independently of us. As we have seen, this is the conception that underlies the use of the power set. The modern literature of positivism, which is very interested in logical matters, usually argues here by reference to Plato. Thus, as Plato believed that only the turn to ideas made men into true men, so it is supposed that only the use of all sets makes mathematicians into true mathematicians. I must admit that I am too much impressed by the undogmatic form of the dialogues to be able to see Plato as a partisan of modern dogmatism. The historical-philosophical connection of the problematic of power sets with Plato is very vague. More accurate historically is its similarity to the problem of universals in scholasticism. The devotees of power sets should be compared with the realists, and the naysayers should be compared with the nominalists. At any rate, the more precise distinctions within scholasticism as well as the distinctions within both realism and nominalism would be for the most part lost in such a comparison. Because I do not intend to try here to resurrect the old metaphysics, we must seek the origin of this difference in conceptions, in its "place and role in life." In my judgment we will find this in the underlying understanding of *language*. A person can already speak before he begins to think scientifically. Thus it is easy enough for a person to understand even his scientific propositions on the basis of his received linguistic framework. Language is not made by man; it is much more that language enables man to create his world. Given poetic expression, this observation becomes: Language is "the house of Being." If one continues to understand mathematics—as was done in the period from Galileo to Kant—in the Pythagorean manner as the language of God or nature and if one therefore carries this understanding of language over into mathematics, then it is not surprising that one would accept that the totality of propositions concerning numbers is given and that by using abstraction one could create the power set. As portrayed here, the "realistic" view of language as the house of Being stands in opposition to the "nominalistic" view, which understands language as a human tool, as an instrument that must first put itself in order. This dispute is difficult to discuss, because discussion is a linguistic phenomenon. Each participant is already speaking from a prior understanding of language. In logical positivism language is viewed as a calculus. That is a very extreme version of nominalism, such as can be found in the modern era only in Hobbes. But even here realism claims its right to existence insofar as the language in which the calculus is discussed, the metalanguage, is always perceived as a language behind which—as with life itself—we cannot go. In my investigations into an operational foundation for mathematics, I have tried to show that it is possible for mathematics to construct languages without having to rely on a prior metalanguage. You can now see from which philosophical wellsprings this effort encounters resistance. Because with linguistic constructions the power set becomes untenable in its so-called uncountability, this effort naturally still has against it the inertia of mathematics as developed to date.

According to the conceptual specification that I have given here for "philosophy of mathematics," all discussion directed toward a substantive decision on these questions should be undertaken by mathematics itself. Therefore, I turn to the third object of my proposed philosophy of mathematics, the preconceptions that underlie the employment of the axiomatic method. First, the employment of this method explains the fact of the unshaken unanimity of mathematical thought, despite the demonstrated dependence upon philosophical conceptions. Mathematics has a thoroughly justified disinclination toward all philosophizing. Its bulwark against philosophizing is precisely the axiomatic method. The oldest part of mathematical science, theoretical geometry, was built axiomatically from the very beginning. Geometry begins with certain propositions, which, as Plato sarcastically observed, the geometer "simply posits as the foundations of his proofs (as if one were completely clear about these propositions) without feeling in any way obliged to give either himself or anyone else a justification, since the propositions are obvious to everyone." Since about 1900 modern mathematics has employed the axiomatic method in all areas. In its claim to be the sole sanctifying method, complete axiomatization has until now remained unshaken by subsequent results that have made clear the limitations of this method. It has been disturbed by neither Goedel's incompleteness result, according to which no consistent axiomatization of arithmetic can produce all arithmetical propositions, nor Skolem's relativity result, according to which an axiom system of set theory, if it is satisfiable at all, is already satisfiable within arithmetic. Hilbert tried to shore up total axiomatization by using metamathematical proofs of freedom from contradiction. To date this program has not been carried through to completion. If at some time in the future a proof of freedom from contradiction should succeed, it can easily be forecast now that it will not achieve Hilbert's goal.

So we still find in philosophy of mathematics the need to inquire into the preconceptions that underlie complete axiomatization. Anyone accustomed to viewing all of mathematics from an axiomatic perspective will look upon this inquiry as meaningless. Use of the axiomatic method will appear to him to be independent of all philosophical conceptions. It appears to be the only possible method, because in order to prove propositions we must already have some propositions from which we can draw inferences. This customary view, which can be traced back to Aristotle, overlooks the fact that we can also begin with constructions instead of with propositions. For example, if numbers are defined as those figures that can be constructed from a single stroke / by the addition of further strokes /, then the proposition that for every number n there is a number $n/$ is not an unprovable axiom; rather, it is a logical consequence of the definition of numbers, that is, a logical consequence of their construction.

This obstinate clinging to axiomatization must have other reasons and causes. In my opinion, it corresponds to the contemporary collective ruling conception of human knowledge. I am inclined to call this conception the "myth of the middle." According to this conception, truth can be compared to a tree. The person who troubles himself concerning truth first grasps this tree about the trunk. That is simple and straightforward. The scientist is that person who then climbs up the tree to follow its limbs as far as they go in all their branchings. Only the philosopher remains below, standing at the trunk, and, from the perspective of the scientist, comes to the absurd idea of digging out the roots of the tree. The tree of this myth is distinguished from real trees in that its limbs as well as its roots stretch to infinity. To that extent truth is, then, not like a human construction, which may indeed have a cellar but in any case has under the cellar an underlying secure foundation upon which everything rests. It is much more that there is no foundation, only a restless effort to travel a finite piece along the infinite stretches that extend above and below. This myth is only of recent provenance and is particularly clear in Pascal, who said in the *Pensées*, "What then is man in the world? . . . a middle term between nothing and everything. What will one do, when we recognize nothing more than . . . what appears of the middle of things, knowing in eternal despair neither the end nor the beginning of things? All things grow out of the nothing and loom into the infinite." For Pascal this dual infinity was a reason to discard science as idle. Today, it is axiomatization, which defiantly contents itself with the appearance of the middle of things, that corresponds to the basic feeling of heroic nihilism. Things have come to the following pass. If the axioms were valid (according to Plato) "as self-evident to everyone" from antiquity to the previous century, then the development of non-Euclidean geometry has shaken this evidence. Since then (that is, for only a hundred years), axioms have been valid as hypotheses that eventually must be replaced by others in the course of extension or application of the theory.

With this conception, axiomatization steps into a close connection with the natural sciences. Hilbert occasionally even dragged in theoretical physics as a justification of mathematics. Physics has stood the test; therefore, the mathematics and therefore the axiomatization of the mathematics used in it must be correct. That is to say, more or less, that mathematicians believe in the axioms because the axioms are supposedly physically indispensable, whereas the physicists believe in these axioms because they are supposedly mathematically secure. If we want to take issue with axiomatization, then we must extend the philosophy of mathematics to a philosophy of physics.

At this point I do not want to take this effort further in that direction. It is clear, however, that in the case of physics my proposed specification of philosophy as critique of preconceptions (for which I used the example of mathematics) would be more like the original form of all Western philosophy, namely, the Socratic elenchus, which, to the anger of the citizenry, unmasked all supposed knowledge and revealed it to be opinion. I like to see this as a symbol of the historical legitimacy of appeals to the word "philosophy." Moreover, our example shows that the analysis of present preconceptions cannot be carried out without recourse to the conceptual world of the supposedly dead metaphysics of Aristotle or Kant. It is therefore appropriate—following a pleasant custom of German speeches—to close with a word from Goethe: "In the natural sciences one cannot properly talk about certain problems without calling in the aid of metaphysics . . . that which was, is, and will be before, during, and after physics."

THIRTEEN

Justifying the Deductive Method

If the primary propositions of arithmetic are first introduced as propositions of constructibility (propositions of deducibility) in logic-free calculi (the rules of which are justified by prearithmetic practices) and if, in addition, logical operators are introduced by rules of attack and defense for dialogues about composite (using logical operators) propositions, then—with additional norms of the entire course of the dialogue (cf. W. Kamlah and Lorenzen, *Logische Propaedeutik* [Mannheim: Bl-Hochschultaschenbücher, 1973])—the propositions of arithmetic are dialogue-definite. That a system Σ of such propositions "arithmetically implies" a further proposition A (symbolically: $\Sigma < A$) is defined by the winnability of the dialogue $\Sigma \| A$ with the hypothesis Σ and thesis A.

The "arithmetic truth" of A is defined by $\forall < A$ for the empty system $\forall$.

In contrast to this "constructive" concept of truth, axiomatics designates a certain system Σ_0 of arithmetic propositions (these are called *axioms*) and also a system of *logical implications*, $\Sigma <_L A$ (fundamental implication). A proposition C is then said to be arithmetically true if, using a chain of logical inferences.

$$\Sigma_0 <_L A_1$$
$$\Sigma_0 <_L A_m$$
$$\Sigma_0, A_1, \ldots, A_m <_L B_1$$
$$\Sigma_0, A_1, \ldots, A_m <_L B_n$$

and so on until

$$\Sigma_0, \ldots <_L C$$

C can be deduced from Σ_0.

This construction of deductive chains is essential to logical inference, that is, the deductive method.

But if in addition to fundamental implication we include in the definition of "logical implication" the following "cut rule"

$$\Sigma_0 <_L A,, \Sigma_0, A <_L B \Rightarrow \Sigma_0 <_L B$$

then—for axiomatization—the truth of an arithmetic proposition C can be defined simply by $\Sigma_0 <_L C$.

A constructivism that seeks to provide a justification of the deductive method must address the following tasks:

1. Define a logical implication in addition to arithmetic implication.
2. Demonstrate the admissibility of the cut rule.

The first task is trivial. We define a position in a dialogue $\Sigma \| A$ to be "logically winnable" when the proponent has only to defend primitive propositions that the opponent has already presupposed. Arithmetic implication $\Sigma < A$ in this sense is valid a fortiori for every logical implication $\Sigma <_L A$. For its justification the deductive method requires only a proof of the metalogical form of modus ponens:

If $< A$ and $A <_L B$, then $< B$

This modus ponens is a special case of the admissibility of the cut rule for arithmetic implication

$$\Sigma < A,, \Sigma, A < B \Rightarrow \Sigma < B$$

The admissibility of this cut rule in constructive arithmetic is nothing but a form of Gentzen's fundamental theorem.

Because the proof of Gentzen's fundamental theorem is still the subject of controversy,[1] I present here a proof as it first appeared (in the context of discussions of general game theory).[2]

The admissibility of the cut rule is proven using a partial formula induction over A.

If it is the case for a primary proposition a that $\Sigma < a$ and $\Sigma, a < B$ and if a is false, then a winning strategy for $\Sigma \| a$ is also a winning strategy for $\Sigma, a \| B$. If a is true, a winning strategy for $\Sigma, a \| B$ immediately gives a winning strategy for $\Sigma \| B$.

For the partial formula induction we now assume that the cut rule is admissible for all partial formulas of a formula A. (If A is quantified,

---

1. Cf. Kleene, *Introduction to Meta-Mathematics* (Amsterdam: Holland, 1952), sec. 79: "To what extent the Gentzen proof can be accepted as securing classical number theory ... is in the present state of affairs a matter for individual judgment."
2. K. Lorenz, *Archiv fuer mathematische Logik und Grundlagenforschung*, 1968.

Justifying the Deductive Method | 173

either $\bigwedge_x A_1(x)$ or $\bigvee_x A_1(x)$

then, for every number n the formula, $A_1(n)$ is called a partial formula.)

If A is a conjunction $A_1 \wedge A_2$ (or the corresponding universal quantification, $\bigwedge_x A_1(x)$) or a conditional $A_1 \to A_2$ (or the corresponding negation, $\neg A_1$), then in each case there immediately follows from $\Sigma < A$

conjunction: $\Sigma < A_i$ (for $i = 1,2$) or $\Sigma < A_1(n)$ (for all n)
conditional: $\Sigma, A_1 < A_2$ or $\Sigma, A_1 < \wedge$ (for the null thesis $\wedge$)

If A is a disjunction $A_1 \vee A_2$ (or the corresponding existential quantification, $\bigvee_x A_1(x)$), we take as a winning strategy for $\Sigma \| B$ a winning strategy for $\Sigma \| A$ up to the position $\Sigma_1 \| A$ in which the proponent defends A, so that we produce the winning position $\Sigma_1 \| A_i$ (for some i) or $\Sigma_1 \| A_1(n)$ (for some n), respectively. It now suffices (because Σ is contained in Σ_1), using $\Sigma_1, A < B$, to deduce $\Sigma_1 < B$. If A is not defended, then $\Sigma < B$ follows trivially.

If we write Σ instead of Σ_1, then we have corresponding to conjunction and conditional for

disjunction: $\Sigma < A_1$ (for some i) or $\Sigma < A_1(n)$ (for some n).

Now, in all cases we use a winning strategy for $\Sigma \| B$ as a winning strategy for $\Sigma, A \| B$ up to positions $\Sigma_2, A \| B_2$, in which the proponent attacks A for the first time. If such an attack does not occur, then we already have a winning strategy for $\Sigma \| B$. Otherwise, it suffices, using $\Sigma_2 < A$, to deduce $\Sigma_2 < B_2$.

For an induction over the number of attacks upon A, we assume that $\Sigma_2 < B_2$ is valid for all strategies in which A is attacked less often than for winning $\Sigma_2, A \| B_2$.

From the opponent's defense against the attack on A, we obtain in the cases of

conjunction: $\Sigma_2, A, A_i < B_2$ (for some i) or $\Sigma_2, A, A_1(n) < B_2$ (for some n)
conditional: $\Sigma_2, A < A_1$ and $\Sigma_2, A, A_2 < B_2$ or $\Sigma_2, A < A_1$
disjunction: $\Sigma_2, A, A_i < B_2$ (for $i = 1,2$) or $\Sigma_2, A, A_1(n) < B_2$ (for all n)

These implications have winning strategies in cases where A is less often attacked. Therefore, using $\Sigma_2 < A$ we can next deduce

conjunction: $\Sigma_2, A_i < B$ (for some i) or $\Sigma_2, A_1(n) < B_2$ (for some n)
conditional: $\Sigma_2 < A_1$ and $\Sigma_2, A_2 < B_2$ or $\Sigma_2 < A_1$
disjunction: $\Sigma_2, A_i < B_2$ (for $i = 1,2$) or $\Sigma_2, A_1(n) < B_2$ (for all n)

Using induction over partial formulas (because Σ is contained in Σ_2, in all cases there then results $\Sigma_2 < B_2$. Only the conditional requires an intermediate step via $\Sigma_2 < A_2$. With that, $\Sigma < B$ is proven for all cases.

With this proof of the admissibility of the cut rule (semantic) consistency has been demonstrated for all axiomatic arithmetics, the axioms of which are constructively true (e.g., the Peano axioms). Every formalized arithmetic that uses a constructive logical calculus (e.g., the Heyting calculus) is thereby proven to be semantically consistent. We know that it is possible to make the transition to classical logic by replacing the disjunction

$$A_1 \vee A_2 \text{ or } \bigvee_x A(x)$$

with

$$\neg\neg A_1 \wedge A_2 \text{ or } \neg\bigwedge_x \neg A(x)$$

All formulas free of disjunctions acquire stability (i.e., $\neg\neg A < A$) constructively from the primary propositions.

To be sure, Goedel's theorem says that an *arithmetization* of the consistency proof presented here leads to a formulation of the assertion of consistency that is not deducible in the Peano formalism, but that is not an objection to the consistency proof; rather, it is only an additional piece of information about the Peano formalism.

FOURTEEN

Pascal's Critique of the Axiomatic Method

The three centuries that have passed since the death of Pascal usually present themselves as dominated by the new method of natural scientific thinking, that is, by a new cooperation between experiment and mathematical deduction.

Pascal would deserve a place of honor in the history of modern science, if only for his unified foundation of aero- and hydrodynamics. But, as regards logico-mathematical thought considered apart from experimentation, Pascal also deserves to be honored. He ensured that the methodological insights of *both* ancient philosophy *and* the Renaissance, rather than the traditional and mystical thought of the Middle Ages, were passed on to the modern era.

Of course, Pascal depended heavily on the work of Descartes, but many of Pascal's formulations are much clearer than Descartes's. In particular, Pascal made clear the role of definition in axiomatic theory. For Pascal, every definition (more precisely, "analytic definition") is only the substitution of a composite expression by another expression and therefore must in principle be eliminable. Descartes and those like Pascal who followed him were champions of the new science, Galilean mechanics—which also means that they fought against the old science, Aristotelian logic and physics.

Today, we are able to do greater justice to Aristotle. Aristotle was the first to clearly formulate the axiomatic method, which was extended to mechanics in the seventeenth century.

1. Human knowledge is built upon certain undefined basic concepts, and all further concepts are defined in terms of these undefined basic concepts.
2. Certain unproven basic propositions, axioms, are valid for the basic concepts, and all further propositions are proven on the basis of these basic propositions.

As regards method in general, there is in Descartes nothing different from what we find in Aristotle. In Descartes the axiomatic method is the sole means for attaining certainty, for attaining indubitable truth.

Now, in my opinion Pascal's remarkable brilliance lies in the fact that at the very beginning of the modern era he had already recognized that modern science's self-confident belief that it possessed a secure method is illusory and a pretension. In this respect Pascal is amazingly modern. However, Pascal was no modern empiricist for whom all axioms are only hypothetically valid and advanced only tentatively. Pascal's critique derives from other sources. As we know, Pascal was very impressed by Montaigne, which is to say that he was impressed by ancient skepticism. But it is also true that Pascal was no skeptic; his critique of Cartesian self-certainty is based more on a novel insight into the finitude of man. And precisely therein lies what is modern in Pascal. Indeed, the finitude of man is the fundamental theme of modern existential philosophy.

In Pascal's case the finitude is expressed by man's existence as a creature lying between two infinities. In fragment 72 of his *Pensées* Pascal wrote:

> For in fact what is man in nature? A Nothing in comparison with the Infinite, an All in comparison with the Nothing, a mean between nothing and everything. Since he is infinitely removed from comprehending the extremes, the end of things and their beginning are hopelessly hidden from him in an impenetrable secret; he is equally incapable of seeing the Nothing from which he was made, and the Infinite in which he is swallowed up. What will he do then but perceive the appearance of the middle of things, in an eternal despair of knowing either their beginning or their end. All things proceed from the Nothing, and are borne toward the Infinite.
>
> If we are well informed, we understand that, as nature has graven her image and that of her Author on all things, they almost all partake of her double infinity. Thus we see that all the sciences are infinite in the extent of their researches. For who doubts that geometry, for instance, has an infinite infinity of problems to solve? They are also infinite in the multitude and fineness of their premises; for it is clear that those which are put forward as ultimate are not self-supporting but are based on others which, again having others for their support, do not permit of finality.

In any event, Pascal's deep insight into the twofold infinity of axiomatic thinking has not been practically developed in the modern era. Pascal's *Pensées* is taken to be only a theological work that is irrelevant for science. For example, Voltaire believed Pascal to be mentally disturbed in his final years, during which his *Pensées* was produced.

It is only from a completely different direction that the finitude of man has been again brought to our attention, so that we can now better understand Pascal. Our modern insight into the finitude of man arises out of reflection upon the method of historical understanding, out of what are called in the English-speaking world the human sciences.

From Schleiermacher to Dilthey reflection on hermeneutics, that is, on the theory of the understanding of human interaction, particularly of spoken and written sentences, has been directed by the following remarkable proposition: "Knowledge cannot go beyond life." Picking up on Dilthey and Husserl, Misch on the one hand and Heidegger on the other have both made clear what it means to say that thinking must begin with life, with the setting within which practical activity occurs. All thinking is a stylization and refinement of that which we are already continually engaged in doing in our practical lives. Pascal articulated this situation within which man finds himself in the following way (fragment 72): "We sail within a vast sphere, ever drifting in uncertainty, driven from end to end. When we think to attach ourselves to any point and to fasten to it, it wavers and leaves us."

Unlike the majority of philosophers since the time of Descartes and Locke, the philosopher who possesses Pascal's preeminently modern insight into the impossibility of going behind life no longer misunderstands himself to be a consciousness that can only first acquire knowledge of the world through sensations, observations, and inferences. For such a philosopher the world is much more an immediate given, lying ready-to-hand. I would like to express this as philosophy's having gained a new immediacy.

That sounds very hopeful. It would be entirely premature, however, to set great store upon this new beginning that arises out of the human sciences. Today, the way of thinking that is oriented on the natural sciences has a very strong and perhaps even still growing influence, even in the sciences that concern themselves with man.

You will now skeptically ask how, then, philosophy proposes to proceed with its putative new immediacy; following what method will philosophy finally, once and for all, achieve solid results?

It must be admitted that not even hermeneutics as a communicable theory is possible in the absence of logical thought or, to put it more generally, in the absence of methodically ordered thought. At this point in the discussion scholars in the human sciences usually appeal to the hermeneutic circle, that is, to the essential circularity of understanding. The search for a methodical starting point for thought is said to be a rationalistic illusion, an illusion (according to this view) in which only the naive positivists with their faith in progress are still caught.

Interestingly, this is not at all the case. Since the 1930s, logical positivism has made use of a conception of thought and speech that ends in an unavoidable circularity—principally in the work of Tarski, but Carnap and Quine have also contributed.

For logicistic philosophy the problem arises in the form of an inquiry about the foundations of scientific language. In particular, the rules of logic are conceived in this inquiry to be the syntactical rules of scientific language. The answer is presented most clearly in a metaphor. According to this metaphor, scientific language with its syntactical rules is a ship upon which we find ourselves—under the restriction that we can never put into harbor. All repairs and additions to the ship must be carried out on the high seas.

In many respects, of course, this picture is correct, but it is explicitly used by logicistic philosophy to cut off any search for a methodical beginning for thought. To be sure, any scientific language, which must be representable as a calculus if it is to be truly scientific, also has a semantics, an interpretation of the symbols of the calculus. For this semantics, however, there must also be available a language called a metalanguage. In practice, natural language functions as this metalanguage—indeed, from the ship of natural language no one can disembark.

At this point and on this issue hermeneutics and logicism coincide. Both schools deny that there can be a methodical construction of thought. To infer from this coincidence that this denial is therefore also necessary would be to infer merely from a fact. It seems to me, faced with this coincidence, that what is really necessary is that we be doubly cautious.

I want first to note that Dilthey's dictum, quoted earlier, that knowledge cannot go beyond life may not be claimed as a proof of the necessity of the denial of a methodical beginning for thought (or knowledge). It is only that this beginning may not be sought beyond life.

Someone will then object that life, the practical situation in which we find ourselves before we begin to engage in science or to philosophize, also includes our use of natural language and its syntax. I admit that, but it does not mean that we would be compelled to put natural language and its rules at the beginning of our *planned* methodical construction.

If we accept this view of natural language as a ship upon the sea, we can describe our situation much better as follows.

If there is no solid ground to be reached, then the ship must have been constructed on the high seas, not by us but by our ancestors. These people must have been able to swim and—somehow, perhaps out of driftwood—they must have first carpentered together a raft and then continually improved that raft until today it has become a ship so comfortable that we no longer have the courage to leap into the water and start once more from the beginning.

As regards the problem of thought, however, we must place ourselves in that shipless (i.e., without language) situation and must try to repeat the actions by which a person can build a raft or even a ship while swimming in the middle of the sea of life. Because such an effort fits neither hermeneutic philosophy nor logistic philosophy, I can no longer appeal to any authority in what follows—unfortunately, not even to Pascal. There is nothing more for me to do but to commence here and now with this construction. I want to sketch here the path that leads to logic and arithmetic. Because I want in no way to presuppose any knowledge of either logical operators or numbers, I must go back a little further than is usual. We must go back to the very first possibilities of thinking and speaking.

One last preliminary observation is necessary, however. You might suspect that I shall now quit speaking English because I do not want to presuppose any metalanguage. That is not what I mean. In what follows, however, I will use the English language only to describe what you would have to do if you wanted to construct a language methodically. This description can be replaced by the sort of practical instruction that is given to children who cannot yet speak. Conversely, what I shall say here can replace such practical instruction.

If in the following I occasionally remind you of certain commonplaces concerning natural language, it will only be to put you, who already know such a language, more quickly in the picture.

I want to start with the established fact that in all natural languages (say, Chinese, Hopi, Ewe, or whatever) that linguists have examined it is possible to construct sentences syntactically. Strictly taken, this point of inquiry is intelligible only if it is directed toward a specific language, but every language has the equivalent of the syntactically simplest sentences, which in English take the form of "This is so" and "This is not so." Sentences of this form may be called *elementary propositions*. Such elementary propositions are in turn constructed using a subject and a predicate. Viewed methodically, these propositions are preceded only by sentences of one word in which the subject is replaced by the situation, a referring gesture, or something similar. Thus only the predicate is expressed in such propositions. We can always introduce new predicates by using a sufficient number of examples and counterexamples. I should like to call this procedure the exemplary introduction of predicates. Which things are to be called "chair" and which not, when something is to be called "clean" and when not, we all learn through exemplary introduction. I am not appealing here to a fact of child psychology (neither is it a matter of a scientific determination of the ontogenesis, or even the phylogenesis, of speech). I want only to remind you that we ourselves are able to introduce predicates exemplarily.

It is obvious that with exemplary introduction the use of predicates is still very imprecise. Nonetheless, this is a possible beginning for speech.

If you use a predicate, then you always intend a particular thing to which this predicate is attributed or denied. This particular thing does not have to have a proper name. For example, normally children learn the predicate "doll" before they give their dolls proper names.

If we have proper names and predicates at our disposal, then we can construct elementary propositions. Subjects are the proper names for particular things, and the predicate is attributed or denied to a particular thing by means of a copula—in English, the copulae are "is" and "is not."

Only propositions that concern a particular thing and that in addition have only one predicate are called elementary propositions. However, it will be possible for several subjects to occur in an elementary proposition as, for example, in the proposition "Max and Moritz are brothers."

The forms taken by elementary propositions are then as follows:

$$\text{affirmative: } S_1, \ldots, S_n \; \varepsilon \; P$$
$$\text{negative: } S_1, \ldots, S_n \; \varepsilon' \; P$$

In every natural language it is possible to predicate a particular thing in this way. Whether this occurs with a copula, as in English, or without a copula is irrelevant for methodical thinking. Every predicate of a language concerns a distinction; the particular thing to which the predicate is attributed is distinguished from the particular things to which the predicate is denied. The predicates introduced exemplarily constitute a system of distinctions that now can serve as a basis for expansion. For this I want to use the term "distinction base."

We now have the task of investigating how we can construct thought from this distinction base using elementary propositions as the available linguistic means. For this purpose, natural languages have available a rich syntax (e.g., logical operators, prefixes, infixes, and suffixes to use for word construction).

If we want to proceed methodically, we must then also acquire such auxilliary linguistics tools. We must construct a rational syntax. We will preclude any appeal to natural syntax. In this way, our project is distinguished from the semantics of logicism, because the latter appeals in the end to the syntax of natural language.

Pascal's Critique of the Axiomatic Method | 181

As a first step leading beyond elementary propositions, it is possible to reduce the imprecision that attaches to the predicates of the distinction base due to their having been introduced exemplarily.

For this next step, which will lead us to concepts, I must give examples if I am to remain true to the method I have suggested. Let us then assume that certain predicates, which in English would be rendered by "living being," "man," "animal," "plant," "raven," have been introduced only exemplarily. The use of these predicates can be made more precise only through rules. You will understand without further elaboration which rules I mean if you look at figure 14.1. These rules for the predicates

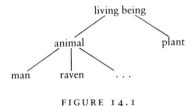

FIGURE 14.1

"living being," "animal," and "plant" can be written explicitly as

$$x \; \varepsilon \; \text{animal} \Rightarrow x \; \varepsilon \; \text{living being}$$
$$x \; \varepsilon \; \text{plant} \Rightarrow x \; \varepsilon \; \text{living being}$$
$$x \; \varepsilon \; \text{animal} \Rightarrow x \; \varepsilon' \; \text{plant}$$
$$x \; \varepsilon \; \text{plant} \Rightarrow x \; \varepsilon' \; \text{animal}$$
$$x \; \varepsilon \; \text{living being}, x \; \varepsilon' \; \text{animal} \Rightarrow x\varepsilon \; \text{plant}$$
$$x \; \varepsilon \; \text{living being}, x \; \varepsilon' \; \text{plant} \Rightarrow x \; \varepsilon \; \text{animal}$$

Such rules are not universal propositions; the nature of such propositions is still unclear at this point. These rules are more practical instructions that prescribe the movement from certain propositions to certain other propositions. The symbol ⇒ (arrow) is introduced *exemplarily* as the expression for such movements. The activities prescribed by these rules can be learned through *practical* training. The rules are no more imperatives or sanctions than they are the rules of a game. Rather, these rules are the building blocks of a language to be learned, a language game, as Wittgenstein said.

We could also order the employment of the predicates given above in other ways—for example, as illustrated in figure 14.2. It is pointless to look for the "true" system of rules before we have employed such rules. We must first begin with some rules or other; only later can reflection lead to suggestions for improvements. Plato effectively compared the

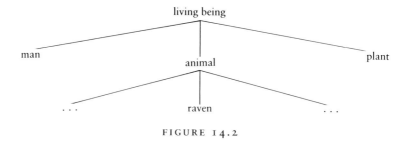

FIGURE 14.2

situation with cooking; a person must first learn to butcher game at the joints.

Before we fall to quarreling over such systems of rules, we must first recognize that in fact it is a matter of rules. In the simplest case, say, $x \, \varepsilon \, \text{man} \Rightarrow x \, \varepsilon' \, \text{animal}$, understanding such rules requires simply having practiced the intended action (e.g., construction of the proposition Hans ε' animal *after* the proposition Hans ε man). These rules are conditional maxims exactly like "If you see Mr. X, please say hello to him for me," for example. The "if/then" that appears here (which corresponds to the symbol $\Rightarrow$ above) is not a logical operator. I am drawing not upon the logic that is contained in natural language, only upon our practical ability to act conditionally. Therefore, I call this "if/then," this $\Rightarrow$, the practical if/then. It belongs to a *practical* language that must be learned before any theory can be learned. By means of such rules the employment of exemplarily introduced predicates is specified more precisely. Further elaboration of these rules draws together ever more predicates, which are thereby also more precisely specified; they then form a system of predicates. There always remains, however, the possibility of altering the current system of rules on the basis of new examples.

No matter how the predicates are bound together into a system by the particular rules chosen, each predicate receives thereby, in addition to its exemplary introduction, a place value in the system. The significance of this needs to be explained methodically. In a system it is possible to use the rules to carry out deductions. For example, in the system given above the proposition "This is a living being" is deducible from the proposition "This is a raven."

(1) $S \, \varepsilon \, \text{raven}$
(2) $S \, \varepsilon \, \text{animal}$
(3) $S \, \varepsilon \, \text{living being}$

If the proposition $S \, \varepsilon \, Q$ is deducible from $S \, \varepsilon \, P$ in a particular system, then we say that Q is deducible from P. This depends only on the rules of the system, not on the exemplary introduction of the predicate. Next, we will abstract from exemplary usage. In complicated systems it can happen that Q is deducible from P and also that P is deducible from Q. P and Q are then said to be equivalent within the system. We also then say that the predicates P and Q "represent the same concept." Talk of concepts is methodically introduced in this way. We call this abstraction. Frequently, concepts are presented as being the meaning of predicates, wherein meaning is conceived in analogy to the reference of proper names. A proper name refers to an object. There is nothing problematic about that, because that is precisely the function of a proper name. However, whether predicates should be said to have references is still a matter of serious discussion. Some people have even resurrected the old dispute between realism and nominalism over universals. In the methodical order presented here, talk of concepts is not a problem. We have already said what it means for two predicates to represent the same concept. We now need to consider how we can speak intelligibly about concepts themselves. For that we must show how we get predicates that apply to concepts.

This is done by next introducing predicates that apply to predicates, that is, predicate predicates. The division of predicates into short predicates and long predicates, for example, can be done exemplarily. We place the predicate that we want to talk about in quotation marks, following the customary convention, and obtain, for example,

$$\text{"long" } \varepsilon \text{ short}$$
$$\text{"predicate predicate" } \varepsilon \text{ long}$$
$$\text{"short" } \varepsilon' \text{ long}$$

We now consider predicate predicates that, if they are valid for a predicate P, are then also valid for every equivalent predicate. Predicate predicates of this kind may be called *invariant*. An example of an invariant predicate predicate is the two-valued predicate "is deducible from."

Now, to speak of concepts means to abstract from all that distinguishes between equivalent predicates, and that means that we *restrict* ourselves to invariant predicate predicates. If R is an invariant predicate predicate, then we now write $|P| \varepsilon R$ instead of "P" εR and read the former as "the concept P is R." The word "concept" indicates here that an invariant predicate predicate of P is being expressed.

This act of restricting ourselves to invariant propositions constitutes the essence of abstraction.

By means of abstraction a system of predicates becomes a system of concepts. The constitutive rules of systems of concepts may therefore be called *conceptual specifications*. The difference between this theory of concepts and the Platonic-Aristotelian conception is essentially twofold. First, as regards concepts it is not a matter of a theory of existents, a matter of ontology; rather, concepts are introduced as something belonging to our activity. They are interpreted not ontologically but operationally. Second, the theory of concepts presented here is not, as in the case of Aristotle, confounded with logic; as the theory of logical operators, logic is a new step yet to be taken.

How can logical operators be methodically introduced into the language constructed thus far? The starting point is again a recursion to the practical situation in which we speak. We think of two people, both of whom use the same system of concepts—for example, the one above—and who engage in a dialogue with each other. What does it mean when one person asserts, for example, "All ravens are living beings"? Such an assertion is definitely completely different from a constitutive rule of a system of concepts.

In a natural language the relevant distinction can be articulated only with difficulty, as long as we are unaware of what the logical operators of the natural language are. In Aristotelian logic our assertion "All ravens are living beings" would be understood as a logical relation between the concepts "raven" and "living being." As remarkable as it sounds, it was only in modern logic (above all, thanks to the investigations of Frege in his *Begriffschrift* of 1879) that it became clear that such an assertion is constructed from elementary propositions using logical operators and indeed in a way that can be formulated in English as

$$\text{For all } x: \text{ if } x \: \varepsilon \text{ raven, then } x \: \varepsilon \text{ living being}$$
$$\bigwedge_x x \: \varepsilon \text{ raven} \rightarrow x \: \varepsilon \text{ living being}$$

The construction is effected here with two logical operators, the junctor "if/ then" and the quantifier (or quantor) "for all." The sense of these operators, however, must still be shown, and for that we must show the employment of such composite assertions in dialogues. To assert a proposition means to take responsibility for defending it against an interlocutor in a dialogue, an opponent. For such assertion and defense to be possible in general, we must lay out how the logical operators are to be employed.

The use of the word "all" in English suggests setting the following employment for the universal quantifier $\bigwedge$: The opponent may choose an x, say, "Hans," and then the first speaker in the dialogue, the proponent of the proposition, must defend the new proposition, Hans ε raven →

Hans ε living being. For this composite proposition we must now fix the employment of the logical operator →. This is done with the help of the practical if/then as follows: If the opponent asserts the if-proposition "Hans ε raven" (and can defend this assertion), then the proponent must assert the then-proposition "Hans ε living being."

Insofar as we have made use here of the logical if/then in one case and the practical if/then in the other, we have avoided a circle or infinite regress from metalanguage to metalanguage.

In the case of the assertion of primitive propositions like "Hans ε raven" and "Hans ε living being" which occur here, we may assume that the partners to the dialogue are agreed on whether the assertions are correctly made or not, whether they are true or not, as we usually say.

In the particular case just discussed, in fact, the proponent will win the dialogue completely independently of whoever this Hans is, that is, independently of which assertions concerning Hans are correct. This is because, if the opponent has asserted "Hans ε raven," then the proponent can defend his assertion "Hans ε living being" merely by appealing to the relevant conceptual specifications. According to these specifications, the proposition "Hans ε living being" is indeed deducible from the proposition "Hans ε raven." We want to say that this deduction is a proof of the assertion. The proof only makes use of the conceptual specifications, as opposed to any knowledge of individual objects like Hans.

Some propositions can even be defended without any appeal to conceptual specifications. A trivial example is "All ravens are ravens" or, more generally, every proposition of the form

$$\bigwedge_x x \, \varepsilon \, P \to x \, \varepsilon \, P$$

The relevant dialogue runs as follows:

Opponent	Proponent
	$\bigwedge_x x \, \varepsilon \, P \to x \, \varepsilon \, P$
? s	$s \, \varepsilon \, P \to s \, \varepsilon \, P$
$s \, \varepsilon \, P$	$s \, \varepsilon \, P$

A proposition that can be defended on the basis of its form alone is called a *logically true proposition*. By "the form of a proposition" we mean the way in which it is constructed using logical operators. I want to give yet another example in which logical junctors also appear. Every proposition of the form $a \vee b \wedge \neg a \to b$ is logically true. This logical truth does not result from that fact that propositions of this form sound self-evident to anyone who understands English; rather, it results from the fact that these propositions in fact always lead to a win in dialogue. The dialogue runs:

Opponent	Proponent
	$a \lor b \land \neg a \to b$
$a \lor b \land \neg a$	L ?
$a \lor b$	?
a | b	R ? | b
$\neg a$	a

Propositions of the form $a \lor \neg a$ will not lead to a win on the basis of their form alone:

Opponent	Proponent
	$a \lor \neg a$
?	a | $\neg a$
? | a	

Of course, this result depends upon the employment of $\lor$ in such a way that we must effectively make a decision in this case between a and $\neg a$. Whoever holds *tertium non datur* to be a logical truth must give to $\lor$ a different sense, but this sense must also stand up to a dialogic employment; otherwise, every dialogue becomes meaningless.

As is well known, the logical truth of any given proposition is not recursively decidable. But the set of logically true propositional forms is countable. This follows easily from the definition. A propositional form is called logically true if there is a winning strategy for it in the dialogue game. We can write out these winning strategies in an appropriate notation. If we then read the steps in the strategies in reverse order, we have a logical calculus, in fact, we have essentially Gentzen's intuitionistic sequence calculus minus the cut rule.

The status of a particular dialogue can always be determined by writing out all of the formulas asserted by, for example, the opponent in their chronological order, plus the proponent's last asserted formula. In this way we obtain the sequence,

Pascal's Critique of the Axiomatic Method | 187

$$A_1, \ldots, A_n \| B$$

The sequence for the above dialogue concerning $a \vee b \wedge \neg a \dashv\vdash b$ appears as in figure 14.3. Both partial dialogues arrive at end points with the fol-

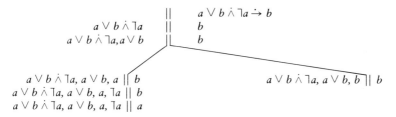

FIGURE 14.3

lowing form.

$$F(c) \| c$$

The preceding sequences result from the "basic sequences" according to specific rules. The explanation and justification of these rules turns on the dialogic employment of the relevant logical operators.

For example, in the dialogic situation $F \| A \frown B$, it is specified that the opponent may question both sides of the conjunction. The proponent must thus be able to defend $F \| A$ and $F \| B$ in order to defend $F \| A \frown B$; that is, we obtain the following rule.

$$F \| A,, F \| B \Rightarrow F \| A \frown B$$

In the dialogic situation $F(A \frown B) \| C$, the proponent can question both sides. To win, it suffices for him to defend one of the two positions, $F(A \frown B), A \| C$ or $F(A \frown B), B \| C$. There results the following rules.

$$F(A \frown B), A \| C \Rightarrow F(A \frown B) \| C$$
$$F(A \frown B), B \| C \Rightarrow F(A \frown B) \| C$$

In a similar way we obtain rules for disjunction $\vee$ and two rules each—one for the proponent and one for the opponent—for the additional logical operators $\rightarrow$ and $\neg$.

For the quantifiers we get the following rules:

$$F \| A(y) \Rightarrow F \| \wedge_x A(x)$$

with y bound in the conclusion, and

$$F[\wedge_x A(x)], A(t) \| B \Rightarrow F[\wedge_x A(x)] \| B$$
$$F \| A(t) \Rightarrow F \| \vee_x A(x) \quad (t = \text{term})$$

$F(\vee_x A(x)), A(y) \parallel B \Rightarrow F(\vee_x A(x)) \parallel B$

with y bound in the conclusion.

We certainly can try to use other operators with other dialogic employments, but these six operators, $\frown, \smile, \rightarrow, \neg, \wedge, \vee$, with the dialogic employment given here (which we could call a synthetic definition of these operators) are at any rate important and useful. The situation seems to me to be similar to the employment of addition and multiplication as elementary operations for calculation. In calculation there are also other operations (e.g., squaring and factoring), but addition and multiplication are in any event useful and important. More than this cannot be said without going into specific applications.

Laying out dialogic employments for the logical operators yields, as we have seen, a sequence calculus that by definition enumerates all logically valid sequences. This calculus does not contain the cut rule

$F_1 \parallel A,, F_2, A \parallel B \Rightarrow F_1, F_2 \parallel B$

and it therefore also does not contain the special case of modus ponens, $\parallel A,, A \parallel B \Rightarrow \parallel B$, although people always take it to be perfectly natural and obvious. In order to explicate the role of this rule, I want to go into the use of logic in mathematical theories. Arithmetic supplies the most important example.

How does one methodically arrive at numbers? We can easily introduce exemplarily words for individual numbers as well as the predicate "number word." We can also add conceptual specifications, which then play the role of hypotheses to be tested. In particular, the infinitude of numbers can be obtained only hypothetically.

The latter situation changes only when we construct number signs; in the simplest form, this is done by following these two rules:

$$\Rightarrow |$$
$$n \Rightarrow n \mid$$

To be sure, in practice it is not possible by following these rules to produce any given amount of number signs, say, 10 to the 100th power. But that is only because life is too short, the supply of chalk too small, or something similar. With these rules any amount of number signs are, as they say, theoretically possible. If a person is not familiar with it, the meaning of the term "theoretically possible" must be taught using just such examples as this.

This construction prescription

Pascal's Critique of the Axiomatic Method | 189

$$\Rightarrow |$$
$$n \Rightarrow n\,|$$

supplies a synthetic definition of the concept of number, as I am inclined to say in a loose allusion to Kant. This prescription also includes an act of abstraction that leads from number signs to the numbers themselves, but I want to pass over that, because the abstraction is carried out in exactly the same manner as in the case of concepts. Calculation operations are also introduced by means of synthetic definitions. We define, for example, how a person is to carry out acts of addition and multiplication:

$$\Rightarrow \frac{m+|}{m\,|} \qquad\qquad \Rightarrow \frac{|\times n}{n}$$
$$\frac{m+n}{p} \Rightarrow \frac{m+n\,|}{p\,|} \qquad \frac{m \times n}{p}\,,\,\frac{p+n}{q} \Rightarrow \frac{m\,|\times n}{q}$$

Finally, arithmetic statements are defined as the logical consequences of these definitions, that is, those propositions that on the basis of these rules (and on the basis of a fixed dialogic employment for the logical operators) can be defended in dialogues against all comers. These propositions may be called arithmetically true and, in distinction from classical arithmetic, to be considered later, may be more precisely called "constructively" arithmetically true.

The true statements of constructive arithmetic are no longer countable. Indeed, constructive arithmetic truth is defined, again, by the existence of a winning strategy. But because the universal quantifier occurs,

$$?n \left\| \begin{array}{l} \bigwedge_x A(x) \\ A(n) \end{array} \right.$$

in which case the opponent can question every possible n (of which there are infinitely many), to win the dialogue with a formula requires the following rule for infinite induction.

$$A(|),\, pA(\|),\, \ldots \Rightarrow A(x)$$

That is to say, a formula $A(x)$ with a free variable x is arithmetically true, if $A(n)$ is arithmetically true for every n. We do not obtain a complete formalism for the deduction of arithmetically true propositions; rather, we obtain only a partial formalism.

For sequences, the partial formalism for constructive arithmetic is as follows.

Basic sequences are:

$$F(p) \parallel q$$

where p and q are elementary propositions for which p is false or q is true.

The logical rules are those in the intuitionistic sequence calculus minus the cut rule.

In addition, we also have infinite induction:

$$S(n) \text{ for all } n \Rightarrow S(x)$$

for any given sequence.

According to these rules, a proposition A for which $\parallel A$ is deducible is winnable in dialogue—and vice versa.

In any case, we must ask what the assertion of deducibility means when made within a partial formalism. A person cannot in general write out the deduction fully. Now, the assertion of deducibility has a dialogic sense. Whoever asserts the deducibility of a sequence must, at the request of the opponent, give a rule according to which the sequence is deducible. The opponent may then choose one of the premises of this rule (there could be infinitely many), and the proponent must then defend this premise and so on until he arrives at the basic sequence.

The formalization that we obtain for constructive arithmetic in this way is essentially distinguished from the standard formalization of intuitionistic arithmetic, which is based on Heyting's work.

In the latter we begin with the Peano axioms. These are formulas that are constructively arithmetically true. Here I only want to show this using the example of the induction schema. The winning strategy looks like this:

$$
\begin{array}{c}
A(|) \frown \bigwedge_x \cdot A(x) \to A(x|) \\
?n \\
A(|) \\
\bigwedge_x A(x) \to A(x|) \\
A(|) \to A(\|) \\
A(\|) \\
A(\|) \to A(\|\|) \\
A(\|\|) \\
\vdots \\
A(n)
\end{array}
\quad \Bigg\| \quad
\begin{array}{cc}
A(|) \frown \bigwedge_x \cdot A(x) \to A(x|). \to \bigwedge_y A(y) & \\
\bigwedge_y A(y) & \\
 & \text{L?} \\
 & \text{R?} \\
A(|) & ?| \\
 & \\
A(\|) & ? \parallel \\
\vdots & \\
A(n) &
\end{array}
$$

Winning strategies for the remaining axioms are also easily given. Finally, the logical implications of the axioms are defined as the statements of the formalized intuitionistic arithmetic, that is, those propositions B for which a sequence

$$A_1, \ldots, A_n \| B,$$

is logically valid, if $A_1, \ldots, A_n$ are arithmetical axioms. In comparing this with constructive arithmetic we find ourselves confronted with the following question. Is the rule

$$\|_{ar} A_1, \ldots, \|_{ar} A_n, A_1, \ldots, A_n \|_{log} B \Rightarrow \|_{ar} B$$

admissible in the partial formalism for constructive arithmetic? Because every sequence that is logically valid is also arithmetically valid, this question is answered affirmatively by the admissibility of the cut rule for the constructive arithmetic partial formalism. This is an extension of Gentzen's principle that appears here in a new light.

This principle justifies the axiomatization of arithmetic. To me, what is important here is that this makes it clear that cut rule elimination must also be demonstrated for the intuitionistic calculus. For the intuitionistic calculus an appeal to intuition does not suffice.

The problem is different in the case of the classical sequence calculus (or any equivalent classical calculus). Here there are several possibilities. We could, for example, use intuitionistic arithmetic's freedom from contradiction to demonstrate classical arithmetic's freedom from contradiction.

We could, however, also use the classical sequence calculus minus the cut rule first to define "arithmetically true formula" for classical arithmetic and then, using Gentzen's principle, directly prove that a partial formalism for classical arithmetic will be free of contradictions.

Even in the latter case (and even if we add an interpretation of truth in classical arithmetic by using, say, semantic specifications or by using a modified dialogic employment of the logical operators) classical logic still has the additional task of specifying precisely its relation to intuitionistic logic, which remains open to interpretation.

One result in this direction is the familiar proposition that every classical arithmetically true formula that contains neither $\vee$, $\bigvee$, nor $\rightarrow$ is also constructively arithmetically true.

With these remarks I hope I have shown how a person can arrive step by step at arithmetic formalism and logical calculi without setting down at the beginning certain unintelligible suppositions (as is the case with the axiomatic method). To do this it was necessary, before discussing logic and arithmetic, that I make some more general observations concerning thinking, in particular concerning predicates, rules, and concepts. This was necessary because the synthetic definitions of the logical operators and the basic concepts of arithmetic make use of the concepts of practical language.

Science has its absolute beginning in practical language—and this (in contrast to the axiomatic method) is completely consistent with man's actual situation as Pascal viewed it: "We sail within a vast sphere, ever drifting in uncertainty, driven from end to end."

PART V

Philosophy of Mathematics: Analysis

FIFTEEN

The Actual–Infinite in Mathematics

Aristotle was the first to distinguish potential infinity from actual infinity—and to banish actual infinity from philosophy and mathematics. The idea of an infinite God, which has its source in Hellenism, was joined—at the latest in Thomas Aquinas—with Aristotle's postulation of the pure actuality of God. Thus arose the Christian notion of God as actual infinity. During the Renaissance, particularly in the work of Bruno, the actual infinity of God was carried over to the world.

The finite world model used today in the natural sciences clearly demonstrates that this domination of thought by actual infinity came to an end with classical (modern) physics. The inclusion of the actually infinite in mathematics, which only explicitly began with G. Cantor around the end of the last century, stands in strange contrast to this. Actual infinity seems anachronistic within the general intellectual and spiritual climate of this century—particularly if we consider existential philosophy.

In what follows I try to show how this anachronism has arisen—and to point out how modern mathematics can get along without this anachronism.

The prototype for all infinities is presented by the cardinal numbers 1, 2, 3, . . . No one would contest the assertion that there are infinitely many cardinal numbers. Russell immediately introduced into logicism (which defines numbers "purely logically" as classes of equivalence classes) an "axiom of infinity" when, despite expectations to the contrary, this infinity was not deducible by "purely logical" means.

Obtaining this infinity is unproblematic, if we introduce numbers not as logical concepts but rather as means for counting. Primitive man used, perhaps, stones or shells for counting. Written signs like I, II, III, . . . are only a more convenient substitute for such "calculi," and our number signs 1, 2, 3, . . . are again only an even more convenient substitute.

We use numbers in counting. But the infinity of numbers is not a consequence of that fact, for how would we ever arrive at infinitely many countable things? The infinity of numbers arises only when we view numbers, which originally are only means for counting, as things in themselves. So we shall not discover anything from a pile of stones. But there is something further in even the primitive form I, II, III, . . . of number signs. We first see this when we realize that numbers can be produced according to a schema. Namely:

1. We begin with I.
2. If we have arrived at x, then we continue by adding xI.

These rules are formulated here symbolically as follows.

(1) I
(2) $x \to x$I

These rules provide a constructive definition of numbers (viz., their construction schema). We can now immediately say that according to these rules infinitely many numbers are possible, for every number x, xI can be constructed. We must bear in mind here that we are asserting only the *possibility*—and this is exactly supported by the rule itself. The rule *makes possible* the construction of xI, if x has already been constructed. It can be that other reasons prevent the construction of xI. The rule itself—as a general rule—is applicable, however, to every x. Anyone who understands this, who has conceived an idea of this construction, has acquired insight into the possibility of this construction.

On the other hand, to assert that infinitely many such numbers really exist, that they can really be constructed by following this rule would of course be false. Aristotle expressed this in Book 3 of his *Physics* as: "In general the infinite exists only in the sense that we always take another and then again another; that thing which is taken is always finite, but in each case a different thing."

In philosophical terminology we say that the infinity of the number series is only potential; that is, it exists only as a possibility but does not actually (i.e., not in reality) exist.

As Brouwer showed in his intuitionistic mathematics, a consequence of this potential infinity is that, for example, the deduction, usually used with finite sets, from a negated universal proposition to an existential proposition [if $A(x)$ is not valid for all x, then there is an x for which $A(x)$ is not valid] may no longer be used. Hilbert's metamathematics then showed that we can nevertheless justify such deductions in the region of the infinite, even if it is only to a certain extent as fictions. It is therefore unnecessary for mathematicians to try to defend such deductions (which are of great importance, because all indirect existence proofs depend on them) using a philosophical argument—namely, the claim that we can *imagine* the set of numbers as an actually infinite set. In any event, there is a construction schema for producing numbers, and therefore their infinity is potential, irrespective of whether we are able always to remember it in our *imagination* or whether we prefer occasionally to forget it. In arithmetic there is—to sum up—no motivation to introduce the actually infinite.

The surprising occurrence of the actually infinite in modern mathematics can be understood only by bringing geometry into our area of inquiry. The previous century—from Cauchy to Weierstrass—produced the arithmetization of geometry. In antiquity it was arithmetic that was geometrized, because the ratios of geometrical quantities could be only partially (i.e., only in commensurable cases) represented in numbers. If we can now represent all ratios using numbers, this is because our concept of number has been expanded. The concept of actual infinity, which in the modern view is instantiated in the real numbers, bears the marks of its descent from geometry.

To understand this connection, we need to look into the problematic of the continuous.

If we have equally spaced points on a straight line,

–|–|–|–|–|–|–|–|–

there is nothing particularly disturbing in this. To be sure, every piece of the line that we can observe contains only finitely many marked points. But if we take a section lying between two such markings and ask ourselves how many points it contains! Now that is very disturbing. It is disturbing because we have before us—so it seems—in a totally unsurveyable form infinitely many points, points so numerous and so thickly packed together that they are inseparable and no longer individually recognizable. Sets of things packed so tightly together we call *continuous*. The paradoxical character of a continuous infinity was already known in antiquity. We find in the fifth century B.C. Zeno's famous "proof" that Achilles can never overtake a tortoise to whom he has given a head start, because in the time that Achilles makes up the head start the tortoise has obtained a new, although also smaller, head start. Should Achilles also make up this head start, the tortoise will nonetheless have again obtained a new one and so on without end. Therefore, says Zeno, Achilles can never overtake the tortoise.

Taking the cavalier attitude that most mathematicians display toward philosophical questions, we could try simply to ignore these "sophistical" problems. Is a continuous length to be identified with the infinite set of its points, or whatever it is in its essence? Why does anyone need to know that when we can constructively deal with length, measure length, and so on? The problem of the continuous, the infinitely small, is more pressing, however, than we may like. In his famous inaugural lecture of 1854 the mathematician B. Riemann said, "The question concerning the unmeasurable is an idle one for natural investigations. But the situation is different when we come to the question concerning the unmeasurably small. The exactitude with which we follow phenomena into the infinitely small is an essential support of our knowledge of the relationships between phenomena." As soon as we move from the measurement of length to the measurement of surfaces we encounter the infinitely small in Cavalieri's well-known principle (Cavalieri was one of Galileo's students). For example, triangles with bases and heights equal are determined to have equal areas in the following way. Every line parallel to the base contains slices of equal length within each triangle (see fig. 15.1). Now, Cavalieri argued, because the triangles are the "sums" of all of their—at least, continuously many—cuts, the triangles have equal areas. I hope that everyone will accept this proof with a certain amount of uneasiness. We can always avoid it by doubling the triangles to obtain parallelograms, which are easily seen to be of equal areas using the right angle produced by the base and height. This solution is not possible when we come to the measurement of space. Someone even proved this impossibility—about fifty years ago. Pyramids with bases and heights equal have equal volumes according to the Cavalierian principle, because every plane parallel to the base contains slices of equal area. The calculation of the volume of a pyramid using the formula

$$\text{volume} = \frac{1}{3} \times \text{area of the base} \times \text{height}$$

can in no way be dismissed as sophistical. This formula was known to the Egyptians, who invented pyramids. And it is at this point that the dissolution of the continuum unavoidably arises; a line is conceived as a set of

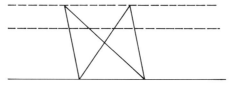

FIGURE 15.1

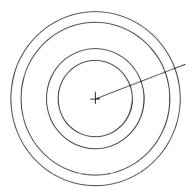

FIGURE 15.2

points, a surface as a set of lines, and a three-dimensional body as a set of surfaces.

To be sure, anyone who knows something of modern infinitesimal calculus knows that in calculus we no longer calculate using the sums of continuously many constituents. For example, we no longer need to agonize over, following the Cavalierian principle, the equal areas of rings of equal thickness (fig. 15.2). Here every line through the center cuts both rings for equal lengths; the area of the rings is reduced to the continuous sum of equal lengths.

But a critical assessment of modern infinitesimal calculus shows that, despite its undeniable and great results, it too "is built on sand," as H. Weyl, one of the most significant modern mathematicians, observed. The use of calculations in the treatment of all geometrical problems, which is customary in the infinitesimal calculus, rests upon the replacement of every line by the set of the line's points. The basis for this replacement is the following connection between arithmetic and geometry. If two points are chosen on a line and the numbers 0 and 1 are assigned to these points, then the other numbers, first integer numbers and then fractions (e.g., -2, $½$, $⅔$, . . .), can also be assigned points on the line (see fig. 15.3).

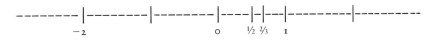

FIGURE 15.3

We normally write fractions as decimal fractions (e.g., $½ = 0.5$; $⅔ = 0.666\ldots$). Now, modern mathematics proceeds on the assumption that every infinite decimal fraction, familiar examples of which are $\sqrt{2} = 1.414\ldots$ and $\pi = 3.14159\ldots$, can be assigned exactly one point. The geometrical line is replaced by the arithmetic set of all real numbers (the name for both finite and infinite decimal fractions). This arithmetic concept of "the set of all real numbers" is apparently geometrically motivated. People also constantly speak of the arithmetic continuum.

The arithmetic continuum is the model for an infinity conceived as actual. There is no schema for construction of the real numbers in Aristotle's sense of taking another and then again another, that thing which is taken, however, always being finite and in each case a different thing. Rather, we imagine the real numbers as actually present all at once; in fact, each real number qua decimal fraction is to be imagined as if the infinitely many digits existed all at once. The radical distinction that exists between finite and infinite decimal fractions is obscured from the outset. A finite decimal fraction can be written down; an infinite one can never be written down. So to speak of an infinite series of digits is, if not total nonsense, at least a perilous presumption. Currently, hardly a word is spent on this issue in mathematical instruction. It is this actual infinity of real numbers (latent in modern mathematics since the seventeenth century) that Cantor first brought explicitly to light, and it is the basis for the present acceptance of the Cantorian conception of infinity.

With the admission of the set of all real numbers as a legitimate object for mathematics there is simultaneously admitted the "power set," that is, the set of all subsets. In the case of a finite set with n members, there are 2^n subsets. In the case of an infinite set, for example, the set C of cardinal numbers, it makes no sense to talk about the set of all subsets of C. To be sure, there are infinitely many subsets (e.g., the set of squared numbers, the set of cubed numbers, the set of numbers taken to the fourth power, etc.), but by what procedure does one obtain *all* subsets?

According to Cantor, there can be no series $M_1, M_2, \ldots, M_n$ that contains all subsets of C, because the set M of those numbers n that are not contained in M is distinct from all members of the series: $M \neq M_n$ (for $n \varepsilon M \leftrightarrow n \varepsilon M_n$.

In connection with this indubitable fact, Cantor created his theory of transfinite numbers, and it was this that led to the first contradictions in set theory.

The "solution" of the problem of these contradictions lay clearly in a prohibition against all actually infinite sets, but this was not acceptable as long as no one knew how to retain the essentials of the classical (modern) infinitesimal calculus. In the nineteenth century the theory of infinitesimal processes like differentiation and integration appeared to rest upon solid ground for the first time, precisely because of the definition of real numbers using "arbitrary" series or sets of rational numbers.

It is clear that we cannot now renounce this classical analysis. A return to Greek mathematics, which was unobjectionably potentialistic, is no longer possible. Greek mathematics was certainly appropriate and suited for its contemporary physics, which was essentially statics and kinematics, but it is totally insufficient for dynamics and the modern field theories of electromagnetism and gravitation as well as for quantum theory. Modern mathematics finds itself in a serious dilemma. In my view, this dilemma is not hopeless. A way can be found to retain the tremendous applicability of modern mathematics on the basis of the potentialistic thesis that the continuum is not an infinite set of points but, rather, only the possible carrier of infinitely many points. Aristotle expressed this in Book 8 of the *Physics* as follows: "In that which is continuous there are indeed, infinitely many points as parts, but these are there not in reality but rather merely as a possibility. For it merely appears that a line has infinitely many points as parts, while by its nature and in its essential being it is otherwise."

The key to the indicated solution lies in replacing Cantor's power set of the set C of all cardinal numbers with an appropriate potentially infinite set. To that end we must specify a procedure for constructing the subsets of C, that is, the sets of cardinal numbers. The possibility of such a construction was first seen by H. Weyl. One needs to proceed on the basis of both of the following facts: (1) Every set of cardinal numbers can be given as the set of numbers x that satisfy a specific propositional form $A(x)$; and (2) propositional forms are finite representations, no matter how varied and numerous the linguistic means may be. Thus, instead of the power set, we have to construct an appropriate potentially infinite set of propositional forms. This task does not have a single solution; there are various constructions that to a greater or lesser degree meet the need. In my "operative mathematics" I have carried out a construction that permits the retention of the principal content of classical analysis essentially unaltered.

If the conception developed here is correct, it represents for modern mathematics a reform, a reform of perhaps the proportion of the introduction of use of the epsilon in the nineteenth century, as the naive concept of continuity was replaced by definitions with ε and δ.

Just as at that time the infinitely small was to be eliminated from mathematics, now also the infinitely large (more precisely, the actually infinite) is to be demonstrated to be dispensable. The driving force behind this reform will be found not in a smug purism but rather in the desire to restore the absolute certainty and unity of mathematics that are today threatened by the set theoretical contradictions and their evasions, using to a certain extent arbitrary set theoretic formalisms.

SIXTEEN

Constructive Foundations of Mathematics

1. THE FOUNDATION OF ELEMENTARY ARITHMETIC AND LOGIC

The ultimate foundation for arithmetic lies in the various prearithmetic practices: the use of signs for counting (such as |, ||, |||, ||||, . . .); the quantitative comparison of signs in place of the quantitative comparison of the things collected; addition and multiplication of signs (in place of certain operations performed on the things collected). These practices *justify*, for example, the following *rules* for the construction of signs, sign-pairs, and sign-triplets:

$$K_1 \begin{cases} \Rightarrow l \\ n \Rightarrow nl \end{cases} \qquad K_< \begin{cases} \Rightarrow l, nl \\ m, n \Rightarrow ml, nl \end{cases}$$

$$K_x \begin{cases} K_+ \begin{cases} \Rightarrow \dfrac{m, l}{m\ l} \\ \dfrac{m, n}{p} \Rightarrow \dfrac{m, nl}{p} \end{cases} \\ \Rightarrow \dfrac{l; n}{n} \\ \dfrac{m; n}{p}, , \dfrac{p, n}{q} \Rightarrow \dfrac{ml, n}{q} \end{cases}$$

Arithmetic begins with *assertions* about the constructibility ($\vdash$) of expressions using these rules. We define, for example,

$$m < n \Leftrightarrow \vdash_< m, n$$
$$m + n = p \Leftrightarrow \vdash_+ \dfrac{m, n}{p}$$
$$m \cdot n = p \Leftrightarrow \vdash_x \dfrac{m, n}{p}$$

To obtain further assertions beyond these primary propositions, we need to add logical operators. As our prelogical practices for these operators we have their employment in *dialogues,* for example, $\neg, \bigwedge_x, \rightarrow$, using the following *rules* for attack and defense:

Assertion	Attack	Defense
$\neg A$	A ?	
$\bigwedge_x A(x)$	n?	$A(n)$
$A \rightarrow B$	A ?	B

These operators alone are sufficient to establish a *complete* system of initial theorems (axioms) for ordering $<$:

$$l < xl$$
$$x < y \rightarrow xl < yl$$
$$xl < yI \rightarrow x < y$$
$$\neg x < l$$
$$\bigwedge_x. A(x) \rightarrow A(xI). \dot\rightarrow A(l) \rightarrow \bigwedge_x A(x)$$

The further construction of arithmetic results from the addition of inductively definable propositions. Logic can be expanded by the addition of additional operators; first, conjunction

$A \wedge B$	L?	A
	R?	B

and then the disjunctions

$A \vee B$	?	A
	?	B
$\bigvee_x A(x)$	?	$A(n)$

A *general* rule of dialogue regulates the course of dialogues about logically complex propositions.

We next need to demonstrate that the winnability of a dialogue about B always follows the winnability of dialogues about A and $A \rightarrow B$. This is the equivalent of Gentzen's fundamental theorem (1936).

If we replace the disjunctions above with the "classical" definitions

$$A \vee B \Leftrightarrow \neg \cdot \neg A \wedge \neg B$$
$$\vee_x A(x) \Leftrightarrow \neg \wedge_x \neg A(x)$$

we obtain classical logic (in particular, we obtain the stability theorem $\neg\neg A \to A$) for *all* arithmetic propositions.

2. THE CONSTRUCTIVE FOUNDATION OF ANALYSIS

The arithmetic of rational numbers makes use of the procedure of *abstraction*. Pairs of cardinal numbers are defined to be "equivalent."

$$m_1,n_1 \sim m_2,n_2 \Leftrightarrow m_1 \cdot n_2 = m_2 \cdot n_1$$

For propositions $A(m,n)$ about pairs that are *invariant* in regard to $\sim$, that is,

$$m_1,n_1 \sim m_2,n_2 \to A(m_1,n_1) \leftrightarrow A(m_2,n_2)$$

we introduce a new notational form,

$$A\left(\frac{m}{n}\right) \Leftrightarrow A\,(m,n)$$

for example,

$$\frac{m}{n} > 1 \Leftrightarrow m > n$$

$$\frac{m_1}{n_1} = \frac{m_2}{n_2} \Leftrightarrow m_1,n_1 \sim m_2,n_2$$

The following is then valid:

$$\frac{m_1}{n_1} = \frac{m_2}{n_2} \leftrightarrow_{A\,inv} A \cdot A\left(\frac{m_1}{n_1}\right) \leftrightarrow A\left(\frac{m_2}{n_2}\right)$$

With this we now have a theory that in addition to cardinal numbers has new *abstract* objects ($r, s, \ldots$), the rational numbers.

Analysis makes use of further abstract objects, principally *functions* and *sets* (of rational numbers).

Rational functions, for example, are abstracted from rational terms.

$$T(r) = \frac{a_0 + a_1 r + a_2 r^2 + \ldots + a_m r^m}{b_0 + b_1 r + b_2 r^2 + \ldots + b_n r^n} \qquad (b_n \neq 0)$$

Terms are said to be "equivalent":

$$T_1(r) \sim T_2(r) \Leftrightarrow \wedge_r T_1(r) = T_2(r)$$

For propositions about terms that are invariant (as regards $\sim$), we write,

$$\bar{A}(\iota r T(r)) \Leftrightarrow A(T(r))$$

Abstract objects $f, g, \ldots$ are called *functions*. For $f = \iota r T(r)$, we define $f\iota r \Leftrightarrow T(r)$, and similarly for many-valued functions.

Sets are abstracted from propositional forms (formulas) $F(r)$. Formulas are said to be "equivalent":

$$F_1(r) \sim F_2(r) \Leftrightarrow \wedge_r . F_1(r) \leftrightarrow F_2(r)$$

For propositions about formulas that are invariant (as regards $\sim$), we write:

$$A(\varepsilon r F(r)) \Leftrightarrow A(F(r))$$

We define, for example, for $M = \varepsilon r F(r)$

$$r_\varepsilon M \Leftrightarrow F(r),$$

and similarly for many-valued sets (relations).

On the basis of this abstraction, the following are valid:

$$\wedge_r . r \varepsilon M_1 \leftrightarrow r \varepsilon M_2 . \rightarrow M_1 = M_2 \text{ (extensionality)}$$
$$\vee_M \wedge_r . r \varepsilon M \leftrightarrow F(r) \text{ (theorem of comprehension)}$$

Here the theorem of comprehension is valid only predicatively, that is, only for definite formulas $F(r)$ (those that contain no set quantifiers).

Analysis, then, is to be built up on the foundation of previous constructions and abstractions.

The following *sequences* are a special case of functions (cardinal numbers as arguments):

$$r_* = \iota n T(n)$$
$$r_n = r_* \iota n$$

Rational sequences are defined in the usual way as "concentrated" and as "equivalent." For invariant propositions about r_*, we write:

$$\bar{A}(\lim r_*) \Leftrightarrow A(r_*)$$

The expressions "$\lim r_*$" are called *real* numbers $\xi, \eta \ldots$

Real sequences ξ_*, that is, $\xi_m = \lim r_{m*}$ are represented by two-valued terms $r_{m,n} = T(m,n)$.

Cauchy's completeness theorem is valid for *definite* terms. Moreover,

> Every nonempty *real* set that is restricted above and has a definite subclass has an upper bound.

(subclass of M is defined by

$$r \, \varepsilon \, U_M \Leftrightarrow \underset{M}{\mathsf{w}} \, \xi r < \xi)$$

Using this form of completeness, classical differential and integral calculus can be provided a foundation with modifications that are only inessential for application (cf. *Differential and Integral* [Austin: Texas University Press, 1972]).

The rest of "modern" analysis remains—to date—without foundation.

SEVENTEEN

Classical Analysis as a Constructive Theory

The controversy over Cantor's set theory has been in a deadlock since the 1930s. With the so-called difficulties of intuitionistic mathematics, formal mathematics, being the simpler of the two, has carried the day. The theoretical foundations of formal mathematics, which Hilbert wanted to give in his metamathematics, have in fact run into difficulties with Goedel's results, but it is precisely these difficulties on both sides that have fostered the development of a pragmatic formalism. Use is made of formalizations of Cantor's set theory, because the latter is supposedly indispensable for classical analysis—and because classical analysis is for its part indispensable for other sciences, particularly for technology and industry, that is, the survival of modern society.

Faced with this "survival value," one stifles all philosophical scruples. It would be unreasonable, however, to call into doubt modern society's right to existence simply because, for example, Aristotle and Kant have claimed that for man infinity is only potential; that is, it can be thought of only as a possibility.

It seems to be profitable to investigate this fundamental mistrust of philosophical speculation and philosophically to criticize the naive self-confidence of modern science upon which that mistrust rests. But that would presuppose that we were once again ready in all calmness and profundity to take up from the beginning (i.e., from Plato and Aristotle) Kant's inquiry in theory of knowledge about the very possibility of non-empirical sciences.

Because I have the impression that this readiness is lacking, at least among mathematicians, I will leave philosophical criticism to the side in the following.

Instead, I propose that as mathematicians we ask whether the transformation that classical analysis has undergone since Cantor, in particular its axiomatization at the turn of the century, could not be replaced by a different transformation that would avoid all the set theoretic difficulties without thereby losing its usefulness for the other sciences. A comparison with classical arithmetic (number theory) would be instructive—I am thinking of Fermat, Euler, and Gauss and of Dedekind's and Peano's transformation of that work into an axiomatic theory. In the case of arithmetic, the basic controversy is in fact a purely philosophical one, because both constructivists and axiomaticists accept the same mathematical theorems; only the question of their interpretation is contested.

The constructivist begins by constructing numbers in which he first constructs number signs, say, in the primitive form |, ||, |||, . . . ; in contrast, the axiomaticist writes down as axioms that there is a number | and that for every number n there is a further number $n|$.

The constructivist then proves the theorem of induction,

$$A(|) \wedge \bigwedge_m . A(m) \to A(m|) . \to \bigwedge_n A(n)$$

perhaps by giving a strategy for defense of this proposition. Although the axiomaticist is certainly able to understand this presentation, he will not allow anything like this as a "proof," preferring to restrict himself to the deduction following certain rules of a logical calculus, the addition of which has no clear motivation.

As you can see, there is not much difference, if we ignore all the words and pay attention only to the formulas that get written down in the argument.

Thus we can say that there is a constructive model for the Peano axiomatization. In fact, it is a constructive arithmetic, and—whether or not the axiomaticists are willing to acknowledge it—that is the reason why constructivists can accept all deductions from the Peano system of axioms as arithmetic propositions.

In the case of classical analysis, the situation is completely different. Suppose we take the axiom system required to construct a complete, Archimedean, ordered structure for the real numbers; then, in the completeness axiom there will appear the concept of an arbitrary "set of real numbers." This concept requires that we then add an axiomatic set theory. Only in this way can we obtain an axiomatic theory that is formulated in the language of elementary logic (i.e., using junctors and quantifiers). The difference here from arithmetic lies in the fact that for such axiomatizations of classical analysis there is no known constructive model.

What is to be done about this situation? We can try to represent as misplaced the inquiry after a constructive model; that would lead us inevitably into philosophy. We could instead try to find a constructive model; there is no objection to that, but I have no idea where I should look for it.

Instead, I want to propose that we turn the point of inquiry upside down and try to find not a constructive model for a given axiomatization but, rather, an expansion of constructive arithmetic, for which afterward anyone who so wishes may try to find an appropriate axiomatization.

In the following, then, I shall sketch a constructive theory that could take the place of a naive classical analysis—and I shall therefore take pains to ensure that this constructive theory is at least as useful to physics as the present axiomatic theory supposedly is.

Because until this century analysis was never an axiomatic theory but, rather, was naively constructive, we should expect that no artificial constructions are necessary and that only constructions intended in the tradition of analysis should be made explicit.

We know that this is the case for the expansion of the arithmetic of natural numbers into an arithmetic of rational numbers and even on into algebraic numbers. Therefore, I shall presuppose here a constructive theory of rational numbers (using an equivalence relation between fractions, i.e., pairs of natural numbers).

We first encounter difficulties when we look for a suitable means of constructing real numbers, because we must simultaneously construct suitable sets and functions for all of the numbers to be constructed. As the goal to be achieved by a reconstruction of classical analysis, I posit the fundamental theorem

$$ID = \Delta$$

For every one-valued differentiable function ϕ, the derivative $D\phi$ is integratable and the integral $ID\phi$ is the difference $\Delta\phi$. At a minimum the known elementary functions should be included in the differentiable functions, and it must be shown that they can be developed in their Taylor's series. For many-valued functions, the fundamental theorem must be generalized to Stokes's theorem. Using differentials, this theorem is written with differentials in the form

$$\int d = \oint$$

That is, for every differentiable differential form ω, $d\omega$ can be integrated, and the integral $\int d\omega$ is the boundary integral $\phi\omega$. As the region of integration, we must at a minimum require all continuously differentiable topological representations of finite sums of intervals.

Hereby we have indicated how far the analysis we are constructing should go. These requirements do not require that the integral to be applied must be Riemannian; on the contrary, the requirements are more easily satisfied using a fully additive integral.

It would require an entire book[1] to carry out this program. Here I want only to detail those areas that require a divergence from the formulations of Cantor's set theory. I give here the *complete* list of these areas of divergence, so that the reader may judge for himself whether the simplicity or usefulness of classical analysis is thereby impaired. Divergences occur in four places:

1. the concepts of set and function
2. the completeness of real numbers
3. some of the theorems concerning real functions
4. some of the theorems concerning multidimensional regions

1. THE CONCEPT OF A SET

Usually the act of abstracting (i.e., the transition from certain given objects, between which there exists an equivalence relation, to abstract objects such that the abstract objects represent the given objects as regards the equivalence) is reduced to the act of forming a set, namely, the formation of equivalence classes. Nonetheless, the reverse is necessary here: the introduction of sets (and functions) as abstract objects.

Non-set-theoretic abstraction occurs as a *façon de parler;* we restrict ourselves to invariant propositions concerning the given objects. The validity of the propositions must be invariant with respect to replacement of a given object by an equivalent object. We then formulate invariant propositions as if they were propositions concerning new (in fact, abstract) objects.

The simplest example is rational functions that take rational numbers as arguments. If r, s, . . . are variables for rational numbers, then let there be constructed rational *terms*

$$\frac{a_0 + a_1 r + a_2 r^2 + \ldots}{b_0 + b_1 r + b_2 r^2 + \ldots}$$

Two terms, $S(r)$ and $T(r)$, are said to be equivalent if $S(r) = T(r)$ for all r. We say of equivalent terms that they describe the same (rational) function—functions are term-abstracta.

Correspondingly, sets of rational numbers are constructed from formulas (i.e., propositional forms).

For fundamental formulas we have, then, term-equations, $S = T$, and

---

1. Lorenzen, *Differential and Integral* (Austin: Texas University Press, 1972).

later we shall acquire additional formulas by using logical operators $\wedge, \vee, \rightarrow, \neg, \bigwedge_r, \bigvee_r$ to combine formulas.

Two formulas $A(r)$ and $B(r)$ are said to be equivalent, if

$$\bigwedge_r . A(r) \rightarrow B(r).$$

We say of equivalent formulas that they represent the same set. We formulate an invariant proposition about the formula $A(r)$ as a proposition about the set $\varepsilon_r A(r)$.

The ε relation is specifically defined by $s \: \varepsilon \: \varepsilon_r A(r) \Leftrightarrow A(s)$. What kind of sets there are depends on our construction of propositional forms. To obtain additional fundamental formulas, let us apply schemata for inductive definitions. For analysis, for example, it suffices to have the following schema for inductive definition of a many-valued set M (i.e., a relation),

$$1, x \: \varepsilon \: M \leftrightarrow A(x)$$
$$n + 1, x \: \varepsilon \: M \leftrightarrow B_M(n, x)$$

in which the x is a variable for any given object. The formula $B_M(n, x)$ may contain only partial formulas $s, T \: \varepsilon \: M$ with $s \leq n$. Under this restriction there is clearly for every n precisely one formula $C(x)$ to be constructed, which is equivalent to $n, x \: \varepsilon \: M$; that is, is defined inductively by the schema.

For analysis it suffices to satisfy these minimal requirements (allowing logical operators and inductive definitions) on the linguistic tools to be constructed. It is not necessary to decide that these complete the construction of a language. Certainly, every completed construction supplies only countably many formulas, but then the procedure, using Cantor's diagonal procedure, can be expanded by means of an inductive definition of an enumeration of all formulas. If we include the power set, say, the power set of the set of natural numbers, among all "possible" sets of natural numbers, there is no linguistic construction that can produce such a power set. The power set is therefore not an object of any constructive theory. Nevertheless, propositions about all its elements, about all sets M of natural numbers, are easily formulated; for example, for all sets M:

$$M : 1 \: \varepsilon \: M \wedge \bigwedge_M . m \: \varepsilon \: M \rightarrow m + 1 \: \varepsilon \: M . \rightarrow \bigwedge_n n \: \varepsilon \: M.$$

The validity of this proposition does not depend on our knowing which linguistic structures are used to represent it. Because the quantifier "for all sets" does not refer to a closed (constructible) range, the quantifier is called an *indefinite* quantifier.

The generality of logical theorems, for example,

$$A \wedge B \to A \vee B$$

is also only formulatable using an indefinite quantifier "for all propositions A, B."

The set theoretic theorem

$$M \cap N \subseteq M \cup N \text{ for all sets } M, N$$

is only a different formulation of this logical theorem.

2. REAL NUMBERS

In defining real numbers we can, following Cauchy, use sequences of rational numbers that are concentrated. A sequence of rational numbers is defined as a function the arguments of which are natural numbers and the values of which are rational. Every sequence must therefore be representable using a term $T(n)$. If we speak of *all* sequences, we are using an indefinite quantifier—but there are supposed to be no indefinite quantifiers in the terms that represent sequences. The terms are supposed to be definite.

If we designate with r_* the sequence the members of which are $r_1, r_2 \ldots$, then r_* is said to be concentrated if the following is valid for it:

$$\bigwedge_\varepsilon \bigvee_N \bigwedge_{n_1, n_2 > N} |r_{n_1} - r_{n_2}| < \varepsilon$$

ε is here a variable for positive rational numbers.

Two concentrated sequences r_*, s_* are said to be equivalent, if the sequence of their difference $r_* - s_*$ is a null sequence. The abstract objects that are produced by applying abstraction to this equivalence relation are called *real* numbers. The real number that is abstracted from r_* is designated by $\lim r_*$.

If $\xi, \eta, \ldots$ are employed as variables for real numbers, then it must be noted that "for all ξ" is now an indefinite quantifier. Such a quantifier may not appear, then, in the term that describes a real number, that is, its concentrated sequence. In the following, in order to distinguish between indefinite quantifiers and definite quantifiers (the range of which is constructible), $\mathbb{A}$ will be written in place of $\bigwedge_\xi$ and, correspondingly, $\mathbb{V}$ in place of $\bigvee_\xi$.

A real sequence ξ_* consists of members $\xi m = \lim r_{m*}$. Thus there must be a double sequence of members r_{mn} represented by a two-valued term

$T(m,n)$. If this term is definite, we also say that the real sequence is definite. Cauchy's convergence criterion is then valid; namely, for every *definite* concentrated sequence of real numbers there is a real number as the limit. The customary proof in fact demonstrates how the limit number can be represented using a *definite* term.

Completeness requires the existence of an upper bound for nonempty, limited sets M of real numbers. If we want to construct a rational sequence r_* with the upper bound as the limit using the customery proof, then we must posit

$$r_n = \frac{T(n)}{n} \text{ with } T(n) = \mu_m \underset{M\xi}{\mathbb{A}} \xi \leq \frac{m}{n}$$

It is clear that an indefinite term appears here. The completeness theorem must therefore be restricted. A simple possibility is the following: Regarding the set M of real numbers we consider the subclass, that is, the set

$$\varepsilon_r \underset{M}{\mathbb{V}} \varepsilon r < \xi.$$

We have before us here an indefinite representation of this subclass. But there may be cases in which this subclass admits of a definite description. Then the subclass itself is said to be definite.

The customary proof then easily supplies the following *completeness theorem:* Every nonempty restricted set with a *definite* subclass has an upper bound.

It remains to be shown in the following that this restriction plays no role in the further development of classical analysis, because the restricting condition that there be a definite subclass is satisfied in the cases that occur.

3. REAL FUNCTIONS

We require of a real function ϕ that it—say, within a given interval—be applicable to *all* real numbers taken as arguments. This can be done, for example, in such a way that a function first is applied to all rational numbers and then on the basis of continuity is expanded to any given argument. This is in consonance with the indefiniteness of the set of all real numbers. In the following we will always presuppose that a function confined to rational arguments is *definite*. Continuity will be defined using the concept of the limit of a function.

We posit

$$\lim_{\xi_0} \phi = \eta_0 \Leftrightarrow \bigwedge_\varepsilon \bigvee_\delta \underset{\xi}{\mathbb{A}} . |\xi - \xi_0| < \delta \rightarrow |\phi\xi - \phi\xi| < \varepsilon$$

Classical Analysis as a Constructive Theory | 215

The right side becomes a definite formula, if the variable is replaced by a variable r for rational numbers. We then write

$$\lim_{r \leadsto \xi_0} \phi r = \eta_0$$

It follows trivially that

$$\lim_{\xi_0} \phi = \eta_0 \to \mathbb{A} \cdot \lim_{\xi_*} \xi_* = \xi_0 \to \lim \phi \xi_* = \eta_0$$

is valid. The usual reversal of this theorem is not possible here, because the proof makes use of the theorem of choice

$$\bigwedge_n \mathsf{W}_\xi A(n,\xi) \to \mathsf{W}_{\xi_*} \bigwedge_n A(n,\xi_n)$$

With the restriction to rational variables, however,

$$\mathbb{A} r_* . \lim r_* = \xi_0 \to \lim \phi r_* = \eta_0 \to \lim_{r \leadsto \xi_0} \phi r = \eta_0$$

This reversal suffices, in the usual way, to prove the uniform continuity of continuous functions in a closed interval $[\alpha|\beta]$, because

$$\mathbb{A}_{[\alpha|\beta]} \xi_0 \lim_{r \leadsto \xi_0} \phi r = \phi \xi_0 \to \phi$$

uniformly continuous in $[\alpha|\beta]$ remains valid. As an application of the completeness theorem, the betweenness theorem may be mentioned. If $\phi a < c < \phi \beta$ is valid, then the *definite* set

$$\varepsilon r \phi r \leq c$$
$$[\alpha|\beta]$$

is formed. It has an upper bound δ, and that is precisely the argument wanted between α and β with $\phi \delta = c$.

The definiteness of the rational restriction is satisfied, particularly for elementary functions. It will suffice to take as elementary fundamental functions power functions, exponential functions, and trigonometric functions, including their inverse functions. The power functions j^n with integer exponents are trivial to define. Rational exponents are obtained using the inverse. In this way, for every real number α we define the exponential function α^j for rational arguments. On the basis of uniform continuity, these functions clearly can be extended to all real arguments. Using the inverse produces a logarithmic function.

It is possible to obtain the trigonometric functions in the following way. For complex numbers we posit

$$i^{\frac{m}{2^n}} = \cos \frac{m}{2^n} + i \sin \frac{m}{2^n}$$

In this way, cosine and sine are defined for all dyadic-rational numbers. The arguments are algebraic numbers. The functions produce the usual trigonometric functions when the right angle is assigned the unit measure of 1 (instead of 90).

Additional functions are obtained from consideration of function sequences ϕ_*, especially power series $\Sigma a_r \xi^r$ with definite sequences of coefficients a_*. Uniform convergent sequences of step functions are important for a modern theory of integration. The limit functions of such sequences have no discontinuities except for jumps; that is, in general there exist right-hand and left-hand limits. These functions may be abbreviated as *jump-continuous*.

As is customary, we obtain the theorem that every jump-continuous function has at most countably many jumps. The presupposed rational definiteness here provides precisely a definite enumeration of the jumps. In the proof that every jump-continuous function is the uniform limit of step functions, one usually makes use of indefinite quantifiers. Indeed, we want—if τ is a variable for step functions—to prove:

$$\bigwedge_\varepsilon \mathbb{W} \bigwedge_{\xi \atop \tau [\alpha|\beta]} | \phi\xi - \tau\xi | < \varepsilon$$

For an arbitrary ε we add

$$A(\xi) \Leftrightarrow \mathbb{W} \bigwedge_{r \atop \tau[\alpha|\xi]} | \phi r - \tau r | < \varepsilon$$

On the basis then of jump continuity

(1) $A(\alpha)$
(2) $\mathbb{A} . \xi < \xi_0 \to A(\xi) . \to \bigvee_\delta A(\xi_0 + \xi)$
   ξ

are valid.

Finally, we must show that $A(\xi)$ for all $\xi\varepsilon[\alpha|\beta]$ is valid. We reason indirectly that the set $\varepsilon_\xi A(\xi)$ otherwise has an upper bound $<\beta$, which easily leads to a contradiction. In order to deduce the existence of the upper bound, $\varepsilon_\xi A(\xi)$ must have a definite subclass, which is not the case given the presence of the indefinite quantifier $\mathbb{W}$. Here we have to modify the proof so that the variable τ is restricted to step functions with rational jumps or with jumps of ϕ. The values are, moreover, rational. (1) and (2) are retained, and the subclass $\varepsilon_\xi A(s)$ thereby becomes definite.

A corresponding modification is also required for the theorem that every function ϕ with restricted variation is a difference of monotonic functions. Regarding a constructive proof, the situation at first looks hopeless. We define

$$\sigma(\xi) = \overline{\underset{\alpha < \alpha_1 < \ldots < \alpha_n < \xi}{\text{fin}}} \Sigma_\nu \, | \phi\alpha_\nu - \phi\alpha_{\nu+1} |$$

in the usual way. The subset

$$\varepsilon_r \underset{\alpha_1, \ldots, \alpha_n}{\mathbb{W}} r < \Sigma_\nu \, | \phi\alpha_\nu - \phi\alpha_\nu + 1 |$$

appears not to be definite. As it turns out, it is nonetheless definite, if we have previously proved that ϕ as a function with restricted variation is jump-continuous (which can be done easily). Therefore, ϕ has at most a definite countable set of jumps and

$$r < \Sigma \, | \phi\alpha_\nu - \phi\alpha_{\nu+1} | \to \underset{\beta_1, \ldots, \beta_n}{\vee} r < \Sigma \, | \phi\beta_\nu - \phi\beta_{\nu+1} |$$

is valid with a definite quantifier, the variables of which are restricted to rational numbers or the jumps of ϕ.

With these resources we can proceed with differentiation and integration as in Dieudonnés's *Modern Foundations of Analysis*. The derivative $D\Phi$ of a function Φ is explained in the usual way. In place of the mean-value theorem we prove the following theorem: Every bound of $D\Phi$ is also a bound of

$$\frac{\Delta\Phi}{\Delta j}$$

This theorem has the advantage of remaining valid when the function Φ is everywhere continuous but only ω-everywhere (i.e., with at most definitely countably many exceptions) differentiable. A function ϕ is said to be integratable, if there exists a continuous function Φ, for which ω-everywhere $D\Phi = \phi$ is valid. The difference $\Delta\Phi$ is then uniquely determined and is called the integral $I\phi$ of ϕ. From this there follows the theorem that every jump-continuous function is integratable.

4. MULTIDIMENSIONAL REGIONS

For a multidimensional integral the choice of the domains for integration is always a bit of a matter of taste. For all applications it is sufficient to restrict oneself to open sets and their closures. The theory of these regions can be developed constructively in the following way. The neighborhoods of N-dimensional space are understood to be open intervals (R, S) with rational end points R, S. A definite sequence of neighborhoods is then given by two definite sequences R_* and S_* of rational points.

Unions of such sequences may be called *open regions* and their closures (i.e., the sets of nonexternal points) *closed regions*.

Although the usual proof that a boundary point is a limit point uses the theorem of choice, it is constructively valid to say that every boundary point of an open region is a limit point.

If compact regions are introduced as restricted closed regions, then the covering theorem is valid. The most important special case runs, "Every definite covering with neighborhoods of a closed interval

$$I \subseteq \cup U_*$$

contains a finite covering."

The proof requires a special consideration, because for $N = 1$ the usual proof produces only

$$\bigvee_n \bigwedge_1 , r \varepsilon U_1 \cup \ldots \cup U_n$$

For $N = 1$ the theorem

$$\bigwedge_1 , r \varepsilon U_1 \cup \ldots \cup U_n \to I \subseteq U_1 \cup \ldots \cup U_n$$

must be proved. For irrational $\xi = \lim r_*$ it in fact follows from the presupposition that a partial sequence of r_* lies in one of

$$U_\nu \, (\nu = 1, \ldots, n)$$

So it follows that

$$\xi \, \varepsilon \, U_\nu$$

because ξ is irrational and U_ν is rational.

For $N > 1$, a simple induction suffices. For the theorem that every function that is continuous in a compact region $\bar{R}$ is uniformly continuous, the covering theorem is useless, because no *definite* covering with neighborhoods of $\bar{R}$ results from the presupposition of continuity. But the indirect proof of

$$\bigvee_\delta \bigwedge_{R,S \cdot B} |R - S| < \delta \to |\phi R - \phi S| < \varepsilon.$$

works for rational points R, S—and that is sufficient. The covering theorem permits the introduction of the fully additive Borel measure for open regions. If for an interval U we define the measure $|U|$ as the product of the lengths of the sides, then for finite interval sums $V = \Sigma U_r$, an additive measure is given by

$$|V| = \Sigma|U_r|$$

For open regions $B = \cup V^*$ with an increasing definite sequence

$$V_1 \subseteq V_2 \subseteq \ldots$$

of finite interval sums,

$$|B| = \lim|V_*|$$

is independent of the choice of a definite sequence of interval sums.

Full additivity is needed when the multidimensional integrals—in order to obtain to Stokes's theorem directly—are defined for alternating differential forms ω by

$$\int \omega = \phi \Omega \text{ with } d\Omega = \omega$$

(where ϕ designates the boundary integral).

Aside from the fact that in general the distinction between definite and indefinite quantifiers must be observed, there are no deviations from the formulations currently required.

It therefore seems to me that we may conclude that Cantor's set theory with its theory of infinite cardinal numbers can be dispensed with in all of *classical* analysis. Now, in *modern* analysis since Cantor, people have thought up many theorems that make explicit use of the infinity of cardinal numbers. Here we should consider whether it would not be an improvement to forswear use of these theorems in the future. In any event, as far as I know, these theorems play no role in application. The "survival value" of constructive mathematics is no less than that of axiomatic mathematics; on the contrary, with constructive mathematics we have the further advantage that a mathematician can *understand* what he does (because—to speak philosophically—he has himself produced the objects of his investigations).

EIGHTEEN

Moral Arguments in Foundational Discussions of Mathematics

To understand the moral problematic as it stands in the contemporary foundational disputes of mathematicians, you do not need to know any more of mathematics than what is provided by a good secondary education. You need to know how in arithmetic we begin to calculate using natural numbers

$$|, ||, |||, \ldots$$

customarily written

$$1, 2, 3, \ldots$$

and how we then proceed to rational numbers (e.g., ¾ or ⁵⁄₇) and real numbers (e.g., written as infinite decimal fractions like 0.333 . . . or 0.101001000 . . .). Finally, we define functions for these numbers (e.g., $f(x) = x$ or $f(x) = \log x$. In addition, it would also be desirable to possess some knowledge of at least a few simple theorems such as (using x, y as variables for real numbers) the theorem

$$\bigwedge_x (x > 0 \rightarrow \bigvee_y y^2 = x)$$

Expressed verbally: "For all real numbers x it is the case that, if x is positive, then the square of at least one real number y is equal to x."

It should also be recalled that mathematical theorems like our example above must always be proved. The question of what constitutes a proof, however, represents the beginning of the foundational problems. Only in the last century were intensive efforts first made to provide proofs of all the theorems that are required for calculation with real numbers, particularly for calculating with differentials and integrals. Since about 1900 these efforts have taken a remarkable direction, mainly through the work of Hilbert and Russell. Following the model of Greek geometry, there were posited *without proof* certain initial theorems, so-called axioms, such that the remaining theorems can be deduced from the former using the rules of logic. Hilbert gave the following formalistic form to this axiomatic method found in ancient geometry in its application to arithmetic. One conceives the axioms are sequences of signs devoid of content, as mere formulas—and the rules of logic are conceived as operational rules that make possible the production of further formulas from the formulas that are axioms. For example, the following theorem can be produced logically from our example theorem

$$\bigwedge_x \bigvee_y (x > 0 \rightarrow y^2 = x)$$

because one of the rules of logic permits the production of a sequence of signs of the form

$$\bigvee_y (A \rightarrow B(y))$$

from a sequence of signs of the form

$$A \rightarrow \bigvee_y B(y)$$

To wit, the logical rule permits one to move the quantifier $\bigvee_y$ outside of the parentheses. The meaning (or content) of signs and formulas can be completely forgotten in this formalistic conception. Only the form matters, that is, only the signs themselves. Further formulas are produced only from previous formulas following previously given rules—nothing beyond that. The construction of mathematical objects—numbers, for example—and the proofs of mathematical theorems are reduced to such *formalisms*, to such systems of rules for operating with formulas.

We can now formulate the foundational dispute between two parties called constructivists and formalists as follows. The formalists assert that only mathematics that has been reduced formalistically belongs to mathematical science; all previous invention and proof are prescientific. The constructivists contest that assertion; they view the *construction* of mathematical objects and the *proofs* that are based on these objects with regard to their content as essential parts of mathematical science. Formalisms are viewed only as a—to be sure, frequently indispensable—resource. This characterization of the point of contention may give the impression that it is a dispute over the words "scientific" and "prescientific"— particularly that the formalists in a rather petty fashion want to deny scientific validity to certain things like construction and content-based proofs that other mathematicians value and hold dear. The situation is, however, not so simple. The problems of constructive proofs of certain theorems concerning real numbers have led to a situation where the formalists operate not only with formulas for which there are constructive proofs (even though these proofs are not acknowledged by the formalists) but also with formulas for which there are no constructive proofs, and thus they operate according to rules that the constructivists with even the best of will cannot interpret as logical rules.

Here I want to call *transcendent* those formalisms that constructivists cannot use as resources in their mathematics, and I want to call *immanent* those formalisms that can be used in constructive mathematics. The transcendent formalisms are considered by constructivists to lie outside of scientific mathematics. The following is an example of this transcendence. The form of the theorem we took as our example above

$$\bigwedge_x \bigvee_y (x > 0 \rightarrow y^2 = x)$$

implies the following theorem

$$\bigvee_f \bigwedge_x (x > 0 \rightarrow f(x)^2 = x)$$

Here f is a variable for a real function. By way of proof we can define $f(x)$ as the positive square root of x (and as 0 for $x < 0$).

From the formalistic perspective, what happens here is that from one formula

$$\bigwedge_x \bigvee_y A(x,y)$$

we deduce another formula

$$\bigvee_f \bigwedge_x A(x,f(x))$$

This deduction rule, the so-called theorem of choice, is constructively valid *only* in special cases, such as the one above, *not* in general. If formalistic mathematics employs the theorem of choice as a general deduction rule, then formalistic mathematics is thereby "transcendent."

It would not be surprising if you now have the impression that the constructivists are the petty ones in the foundational disputes. They want the formalisms (which according to constructivists' own criteria—and these criteria are considered prescientific by the formalists—are said to be transcendent) to be excluded from mathematics as *un*scientific. So the constructivists are said to be *pre*scientific, whereas the formalists are *un*scientific. Of the two parties, each holds that the other is petty, whereas they themselves of course are of noble intentions; indeed, it is only for the sake of science that each criticizes the other. The remonstrance of the formalists that their enemies consort with *pre*scientific constructions is usually not meant in a moral sense; the constructivists are felt rather to be intellectually limited—they do not have the breadth of vision to recognize the new possibilities of transcendent formalism. It is the constructivists who bring a moral aspect into the foundational dispute. Namely, they answer that they are not so limited as not to understand that one can also operate according to the rules of transcendent formalism—but this operation is (due to the transcendence) mere frivolity. To put it more clearly, there is no moral justification for a mathematician's wasting his time in this way rather than restricting himself to constructive mathematics using immanent formalisms when formalism is appropriate.

The moral remonstrance will be discussed in the following. With the desolate condition in which moral philosophy finds itself today (to be sure, the present situation has behind it more than 2,000 years of history in which theology and physics have played an apparently unfortunate role), it is easy to understand that the first reaction of formalists to this sally is to avoid the area of morals and ethics. One makes reference to the fact that all science is value-free—the free play of the intellect—and that the constructivists are spoilsports. Obviously, such arguments are the point at which the formalists enter into the realm of moral philosophy. Therefore, before discussing the underlying moral problematic I will briefly note the ways in which morals can, in fact, be avoided. The formalists must—and this is often done—deny the distinction between immanent and transcendent formalisms. Either one denies in general that formalisms are preceded by anything (be it constructions or whatever else constitutes content), or one claims that the formalisms that the constructivists call transcendent are indeed formalizations of contents but nonconstructive contents, so-called intuitions or ideas. Here one arrives then, in fact, at epistemological assertions free of moral consideration—for example, that there are certain nonconstructive ideas. Questions of theory of knowledge are not, however, the subject of this discussion. I do not deny the distinction between immanent and transcendent formalisms, but there exist other possibilities for coming to some agreement—even when one continues to avoid moral considerations. That is, when for certain formalisms there exists *at the moment* no application from the constructive perspective, that does not mean that *in the future* an applicability for constructive mathematics will not be proven. That means that one could conjecture the possibility of a future immanence. At the same moment that one justifies occupation with a formalism by the *conjecture* that this formalism will prove itself to be an immanent formalism, moral considerations are once again left behind. The application of transcendent formalisms is described as a provisional arrangement, and one can confidently entrust to personal judgment whether one will occupy oneself only with formalisms the meaning of which is secure or whether one will also occupy onself with those formalisms in which there is some risk of absurdity. Of course, the risk should not be too great; the judgment of the magnitude of the risk must depend not on moral considerations but, rather, on substance, on mathematical content. Unfortunately, this formula for agreement has so far found scarcely any acceptance. Instead, there rules a kind of holy war between constructivists and formalists in which various more or less disguised moral arguments are employed. In order to discuss this situation I assume in the following that certain formalisms are recognized to be immanent and certain others to be transcendent. The question of which extramathematical considerations can be adduced as moral justification for occupation with transcendent formalisms will be our exclusive concern. I will discuss four arguments under the headings of *custom, necessity, beauty,* and *pleasure.*

The argument from custom is the simplest. The formalists can appeal to the fact that since the end of the previous century transcendent formalisms have been used by mathematicians. It is true that certain words are often used in place of mathematical symbols in areas where formalizations are still not used (this was particularly so before Hilbert and is still the case today wherever Cantor's set theory, which first appeared in 1872, is used), but many of the customary rules of operation are certainly transcendent. Even if it is not exactly old, this tradition of about a hundred years is also the dominant doctrine. By my estimate, today at most 1 percent of professional mathematicians are prepared to forgo the use of all transcendent rules. Why should they when the others will not? Should an activity like the production of formulas according to certain rules suddenly become immoral—formulas that the most famous mathematicians of the last two or three generations have developed and that are used with virtuosity by today's most famous mathematicians? And should one forgo these simply because the constructivists claim that this activity is transcendent? Understandably, the majority are happy to follow the conservative advice simply to ignore the constructivists who criticize the tradition. One can then continue as usual. You cannot argue with someone who has already decided to ignore the arguments you want to make. And, although you cannot argue with the conservatives, it is not necessary to do so, because reference to the mere fact of a custom does not supply any justificatory grounds for the custom.

Very frequently, the fact that certain transcendent formalisms have become established in the tradition is used to conclude that these formalisms are *necessary*—necessary precisely in the sense that only through the use of these formalisms (their application in physics and thereby in technology and industry) are certain human needs met. This is the argument of the pragmatists. Certain formalisms are useful. More precisely, our vital needs require their employment, completely independent of whether or not they admit of a constructive interpretation. Such a need for transcendent formalisms in physics *cannot* of course be inferred from the history of mathematics and physics. No one knows whether or not a historical employment of immanent formalisms would have produced a still better physics. Only if one could name a specific result of physics that can be achieved using transcendent but not immanent formalisms would this argument have force. So far as I know, however, there are no such results. It is again only the strength of custom that speaks for not restricting physics to constructive mathematical resources.

Although the arguments from custom and necessity provide the strongest themes of the formalists, they are weak as arguments. The third argument, the argument of the aesthetes, appears to be stronger; transcendent formalisms are *more beautiful* than immanent formalisms. Today, when most intellectuals are much more interested in aesthetics than in ethics, this argument is sure to find majority approval. Anyone who knows mathematics in some detail knows that within mathematics there are beautiful and less beautiful theorems, elegant and less elegant proofs. We need not pause to discuss this. But the aesthetic argument is supposed to secure a place in mathematics for certain formalisms—formalisms that the constructivists object to as *un*scientific and that are thus not clearly part of mathematics—merely by the fact that it is asserted that they are beautiful. Here it is not a question of judging, regarding a goal acknowledged to be necessary, whether the formalisms are more or less beautiful. Rather, the issue is whether beauty *alone,* independent of any necessity, is sufficient to prove certain formalisms to be scientific.

As you can see, the distinction between art and science is lost in such arguments. In the case of immanent formalisms we can *additionally* speak of their beauty, because they already belong to mathematics by virtue of their necessity. But in the case of transcendent formalisms it is a matter of arbitrary objects for which no necessity is visible so far. Although we want to call some of them beautiful and others ugly, there is lacking any criterion for this; it can only be a matter of personal opinion, at most of group opinion, that is, a matter of fashion.

In fact, you can frequently hear it said that the decision as to which formalisms are beautiful objects for investigation by mathematicians is left to the *tastes* of the mathematician himself. It is true that the decision as to whether a formalism is applicable in constructive mathematics, that is, whether it is immanent or transcendent, can be made only by mathematicians—not according to taste, however, but by *proof* of applicability within mathematics.

If we reject the group taste in this connection, then we come to the fourth and last argument, the argument made by the hedonists that in mathematics one may investigate those formalisms the investigation of which gives the mathematicians the most *pleasure*. This hedonistic argument is frequently mixed with the aesthetic argument. Namely, one explains that one takes pleasure in a certain formalism because it is beautiful, while at the same time personal pleasure or that of the group constitutes the single proof of this beauty. Here we encounter the hedonistic ideology of modern art, according to which the beautiful is that which produces a sensual enjoyment (through the eye or ear). If you restrict yourself to asserting private enjoyment (of either an individual or a group) of a certain formalism, then you assert a mere fact. But, if enjoyment of formalisms is claimed to be not *only* a private interest but *also* a public interest, then mathematics must be compared on an equal footing with painting or music as well as the art of the perfume manufacturer. Would it not be more in the public interest to promote the manufacture of perfume than that of mathematics, as many more people take pleasure in beautiful odors than take pleasure in beautiful formalisms? We would be compelled to this conclusion, if we argue *only* on hedonistic grounds for transcendent formalisms. In reality, however, the various arguments merge together into a murky mass for the formalists.

One always appeals first to custom, that is, to tradition and the dominant opinion; more precisely, one does not argue here at all, because one offers no basis for one's position. If a person wants seriously to try to justify his occupation with transcendent formalisms, then it is advisable, following the model set by Aristotle, first to investigate the need—that is, in the final analysis—of the preservation of human life. Only when this attempt fails would one try to defend formalisms on aesthetic grounds. We have here no other choice, if we take the term "beautiful" in Aristotle's sense, in which he included all that which can be justified without reference to vital necessity. Because the beauty of transcendent formalisms is not to be justified as an embellishment upon necessity, one is reduced to pleasure, to individual or collective satisfaction. But with that one has also come to the end of all argumentation, because there can be no argumentation over taste wherever taste appears as private enjoyment.

Only the number of private interests is relevant in determining the public interest.

These considerations concerning the possible moral arguments involved in the foundational dispute in mathematics provide no decision on these matters. They seem to me, however, to be useful for moral philosophy. However the foundational dispute turns out, it demonstrates that not even the most rigorous of sciences can be provided with a foundation without moral arguments. Even in mathematics one cannot fall back upon "value-free" formalisms—because the *choice* of the formalisms must be given some foundation. The foundational dispute also shows that the significance of formalisms can be secured only by reference back to our own constructions (beginning with the construction of numbers). But constructions are actions. These actions must be defined using constructive prescriptions, that is, using *norms*. And norms must be justified; one must show that it is better to follow the norm than not to follow it. And here we also come finally upon moral considerations, upon the question of the good, that is, of better and worse.

That things do not stand well with moral philosophy at present is also a reason why the foundational dispute of the mathematicians, with its many defective moral arguments, is a telling example. Only on the basis of such concrete examples can a person try—as in Aristotle's time—to *practice* the art of serious moral argumentation and thereby to reappropriate moral philosophy as the doctrine of the *forms* of moral argumentation.

PART VI
Philosophy of Physics

NINETEEN

How Is Objectivity in Physics Possible?

According to current understanding, physics is an objective science, because its task is to investigate "natural laws"—to use the word "objective" with the meaning it had before it became philosophically overworked. That there is such a thing as a natural law is thereby conceived as a fact that a person can only acknowledge, a fact that can neither be "understood" as possible nor "explained."

Historically, this fact has been recognized only since Newton. Only since Newton has it been possible for a person to point to certain physical propositions (particularly the law of gravity) when asked about "natural laws."

But, if the existence of natural laws can be demonstrated only by the presentation of "physical propositions," then we must ask by way of critique: (1) Which propositions belong to physics, and (2) which physical propositions are "laws of nature"?

As regards an answer to the first question, the names "physics" and "natural philosophy" are entirely misleading. "Physics" etymologically denotes that which grows, and, more specifically, "nature" originally denoted only that which is born.

But "physics" in our sense definitely denotes a science of the inanimate. Thus, for a definition of physics, we must first define "animate things." This necessarily leads us ever and again in circles, so long as we do not know with which concepts we must begin. The sole possibility of forming first concepts, which individual sciences take as their respective bases, is to follow Aristotle's example and orient ourselves by means of simple examples. We will take the following as representatives of all that exists: men, animals, plants, and stones.

We distinguish men from animals on the basis that only men have spirit (reason), animals from plants on the basis that only animals have souls (consciousness), and plants from stones on the basis that only plants have life. For stones there is left only what is also valid of the others, namely, that they are *bodies*.

What we have before us here is—to use specialized terminology—an ontology, and indeed—to use a special metaphor—a division into spheres: the sphere of somata, that is, bodies, and—to use the usual hierarchization—*above* this the biosphere, then the psychosphere, and finally the noosphere.

Somatology is the theory of bodies insofar as they are nothing more than bodies. Physics is a part of somatology.

As regards the definition of physics, we need another distinction. Such sciences as geology, meteorology, and astronomy, which concern themselves with individual and particular bodies (e.g., the earth, the earthly atmosphere, or the planets), are not part of physics; they are, rather, individual somatologies.

That the planets move around the sun in approximate ellipses is not a natural law; it is a fact. Only the propositions of physics that serve to explain this fact are called natural laws. The movement of the planets is the classic example of a precise behavioral constant of nature. Were the planets to alter their behavior, that would not be a violation of natural law; rather, we would have learned only that our previous physical explanation of their behavior was insufficient.

Physics does not concern individual bodies; rather, it concerns bodies in general. Physics is general somatology. It can be systematically divided into the theory of the constitution of bodies from parts (this leads on to chemistry and atomic physics, which are indeed closely related) and into the theory of the movement of bodies *without* change in their constitution. The latter part is essentially what we call classical physics. It defines "forces" as the cause of changes in movement.

The general theory of alteration in the movement of bodies effected by arbitrary forces is called mechanics. We explain the changes in movement that occur in inorganic matter (i.e., those that occur without the action of living beings, e.g., men) by assuming special *fields of force*—indeed, by assuming gravitational fields and electromagnetic fields. With this division of mechanics, gravitational theory and electromagnetics in fact compromise the whole of classical physics.

As regards the question "What makes natural laws possible?" I think it should be conceived as an inquiry concerning the possibility of classical physics, in the sense of a general theory of the movement of bodies insofar as they are nothing more than bodies.

Before becoming involved in an answer to this question, I want to point out that this classical physics is something other than the physics that

Goethe was aiming at with his color theory. Classical physics does not deal with the phenomenon of the light that we perceive with our eyes; that is handled by the part of electrodynamics traditionally called *optics*. Similarly, instead of an acoustics concerned with the phenomenon of sound, particularly music, classical physics teaches only an "acoustics" that investigates wave motion in gaseous bodies. Heat appears only as molecular motion. The phenomena of smell and taste appear not at all; these are investigated by applied chemistry.

The classical physics that has existed since Galileo and Newton is not a phenomenological physics, unlike the ancient physics whose name the former has taken over.

You can see this simply in the fact that the propositions of classical physics are mathematical formulas and not propositions of natural language, which we use to speak of perceptions. How is it possible that certain mathematical formulas should be taken to be laws of nature? It is this question that we must answer.

To that end I need briefly to indicate what mathematics is. In this I can restrict myself to that part of mathematics that is used in classical physics, leaving the rest to the side. The mathematical language of physics is supplied by arithmetic, including the infinitesimal calculus. In arithmetic we operate with self-generated symbols, namely, the number symbols that were invented in a well-known fashion expressly for the purpose of counting. Addition and multiplication are the most important arithmetic operations.

Curiously, for the most part, only multiplication appears in physical propositions. I offer the following examples.

the volume of a tetrahedron:
$V = (1/3) A \times H$
(V = volume, A = area of the base, H = height)

distance as a function of acceleration:
$d = (1/2) a \times t$
(d = distance, a = acceleration, t = time)

conservation of momentum:
$M_1 \times v_1 = M_2 \times v_2$
(M = mass, v = velocity)

gravity:
$g \sim M/r$
(g = acceleration of gravity, M = mass, r = distance)

These examples are drawn from various subdisciplines. The volume formula belongs to geometry (Eudoxus), the acceleration formula to kinematics (Nicole Oresme), and conservation of momentum to mechanics (Huygens). Newton's law of gravity belongs naturally to gravitational theory. It is not part of mechanics, as I have here delineated mechanics, because it deals with a special field, not with arbitrary forces.

None of the examples above is part of arithmetic; it is only that arithmetic is constantly presupposed. Previously, I provisionally characterized these propositions as *physical*. There is, however, a problem with this. In our usual way of speaking, we count geometry as part of mathematics, not of physics. Because geometry is not part of arithmetic, this looks like a verbal disagreement. But behind this verbal dispute lies the following question: Does geometry (like the law of gravity) underlie our investigations of experience, that is, our measurements, or is geometry experimentally unrefutable, as is the case with arithmetic?

In Kantian terms the question is whether geometry is an a priori science or an empirical (a posteriori) science. Since around the middle of the previous century the dominant view has been that geometry is not a priori. A fortiori, according to this view, kinematics and mechanics are not a priori. For example, the claim is that the volume formula is an experimental proposition and thus a genuine natural law, not a human production.

According to Kant, on the contrary, even the law of gravitation is a priori. To me the truth appears to lie somewhere in the middle. Mechanics is a priori, whereas gravitational theory is empirical.

The basis for my opinion can only be sketched in bold strokes here. I call your attention to the fact that the sequence of examples above is not only historical but also systematic. In each example the concepts of the preceding example are presupposed, but with the addition of a new concept. The only exception to this is the law of gravitation.

In geometry we deal only with spatial measurements: length, area, and volume. Kinematics adds to this the measurement of time. Acceleration is defined as a measure of the change of velocity; the velocity itself is a measure of spatial change. Mathematically, we are talking here of differential quotients, and the acceleration formula is a mathematical consequence of the definitions.

With the conservation of momentum we add the measurement of mass. Only then are we able to define a measure for force—according to Newton, the product of mass and effective acceleration.

But the law of gravitation neither introduces a new physical quantity of measurement nor results as a mathematical consequence of previous definitions. It is a natural law. We can now say more exactly what that means. According to the law of gravitation an inertial mass M has a gravitational field that effects an acceleration g on every body at a distance r, an acceleration that is proportional to M/r. All of the quantities that appear here are measurable. (The distance r can be measured geometrically, the acceleration can be measured kinematically, and the mass M can be measured mechanically.) Because the law $g \sim M/r$ says something about *all* gravitational fields, it cannot be verified simply through measurement (because there are infinitely many such measurements). It is not even the case that this law can be refuted simply by producing measurements that give clearly divergent results. It is always possible to explain divergent measurements as the results of disturbances. For example, a medium such as the atmosphere can disturb gravitational effects, as can other forces such as electromagnetic forces. The law of gravitation is an empirical generalization and serves as a hypothesis for explaining the phenomena of motion. In physics such a hypothesis is taken as a theorem, if the hypothesis, together with other physical theorems, succeeds in explaining the phenomena of motion. These confirmed physical theorems are our natural laws.

According to prevailing opinion, so-called positivism, *all* theorems of the exact sciences, insofar as they do not belong to mathematics (i.e., to logic or arithmetic), have the hypothetical character of empirical generalizations.

In contrast, it seems to me that the essence of the Kantian theory lies in the fact that, on the one hand, the hypothetical character of natural laws is acknowledged (in contrast to rationalism or realism, which conceives natural laws as if they could reveal to us how things really are—these days we would say, as if they could describe reality), whereas, on the other hand, the domain composed of geometry, kinematics, and mechanics, which lies between mathematics and empirical physics, is established as a nonhypothetical theory of the concepts of space, time, and mass.

I would like to give the name *protophysics* to this domain, which from the perspective of mathematics contains the first steps that we must take before we can perform physical measurements.

Of course, we have accomplished nothing by merely asserting the a priori character of protophysics. Like Kant, we must ask what makes an a priori protophysics possible. Unfortunately, the Kantian investigations of this question are completely unsatisfactory in their details. Most physicists and mathematicians fail to see any problem here—or even dogmatically reject the possibility of the problem. Hugo Dingler and Hermann Weyl are recent exceptions. Many specific questions remain to be clarified, but perhaps some hints will be sufficient to understand the possibility of an a priori protophysics.

As is usual, the first steps present the greatest difficulty. Geometry occupies the central position in protophysics. Our example of the volume of a tetrahedron is found in Book 12 of Euclid but was known to the Egyptians prior to Euclid. We can also check it empirically by comparing, say, the water displaced by three tetrahedrons with that displaced by a rectangular solid of equal base area and height ($V = A \times H$). Euclid, however, gives a proof; moreover, it is definitely a nontrivial proof, because it presupposes all of Book 5. A proof leads from propositions that we already have to a new proposition. So in geometry we must inquire after the very first propositions. These very first propositions are called axioms.

What makes natural laws possible? This question has led us to the question: What makes an *a priori* protophysics possible? Now we have come to the question: What makes geometrical axioms possible?

We are doing geometry when we pay attention not to the material of which bodies consist but rather to the form that bodies take. That is an ordinary, everyday distinction, the importance of which was first seen by Aristotle. As regards form, we distinguish—again, just as ordinarily—among surfaces, lines, and points.

Such natural language distinctions never by themselves provided the foundation for an exact science. To such distinctions we must add what Plato called an idea and Kant called a pure intuition. What this is exactly must be demonstrated using examples. An especially important first example that presupposes no other examples is that of a flat surface. Our natural language more or less tells us when a surface is "flat." Flatness, however, is something more than any previous concepts, because we can meaningfully speak of complete flatness, although we can point to no completely flat surface on a real body.

This puzzling aspect of Platonic ideas is explained by the fact that complete flatness is a precisely formulatable ideal requirement that is only approximately but ever more closely realizable.

The ideal requirement that we place upon a completely flat surface is the requirement of homogeneity: Everything that is valid for one point on a flat surface is also valid for every other point. More precisely expressed: If a geometrical proposition $P(S,P)$ is valid for a surface S and one of its points P, then for every other point P' of this surface the corresponding proposition $P(S,P')$ is valid. If we write $P \varepsilon S$, $P' \varepsilon S$, respectively, for "P is a point of S" and "P' is a point of S," then we obtain the following as an exact formulation of our idealization:

$$P \;\varepsilon\; S \wedge P' \;\varepsilon\; S \wedge A(S,P) \rightarrow A(S,P')$$

That is my first example of an axiom. It is meaningless to suggest here the possibility that the validity of this axiom could be refuted using physical measurement. Physical measurement presupposes that we are able to measure space, time, and mass. Measurement must also be so determined using ideal requirements that we can speak of complete measurements that can be only approximately but ever more closely realized. *Only insofar as there is agreement concerning the ideal requirements placed upon measurement is the objectivity of measurement made possible.* By "objectivity" I mean so-called intersubjectivity; measurement results are independent of individual circumstances, that is, persons. The idealizations by which complete measurement is specified are propositions that can serve as axioms in protophysical theories. All other propositions of these theories are logico-arithmetic consequences inferred from the axioms.

That we can obtain in this way an axiom system sufficient for protophysics (i.e., for geometry, kinematics, and mechanics) has not yet been demonstrated in detail.

Nonetheless, I feel justified in presenting this much of this program to produce a foundation for protophysics, because in my opinion it is the sole possibility for finally understanding—on the basis of such a well-founded protophysics—how objective knowledge of natural laws is possible. To that end I want to summarize once more the necessary steps:

1. Definition of the measurement of space, time, and mass using a system of idealizations
2. Protophysics as the system of logico-arithmetic consequences drawn from these idealizations
3. Objective measurement of the phenomena of motion and formulation of hypotheses concerning force fields
4. Confirmation of hypotheses that explain the behavioral regularities of nature.

What we call natural laws are thus shown to be the results that we acquire when we measure the behavioral regularities of nature using the objective measurements supplied by protophysics.

TWENTY

Constructive and Axiomatic Methods

Mathematics is like a building with many apartments. At a minimum we have arithmetic, analysis, algebra, and topology, as well as geometry and probability theory. The people who live in these different apartments frequently do not understand each other.

The Bourbaki Program held out the promise of a new unity for mathematics insofar as it allows Hilbert's axiomatic method as the sole genuine method for mathematics.

The program was very successful, with the one exception that the "axiomatic foundations" of set theory remain shrouded in darkness.

But, if we ignore this difficulty (since around 1900 mathematicians have learned to live with it), then mathematicians have no fundamental problems with the axiomatic method. There are no problems in arithmetic, because there is no non-Peano arithmetic. There are no problems in geometry, because mathematicians believe that *physicists* can tell them whether space is Euclidean or non-Euclidean. Finally, there are no fundamental problems in probability theory, because there is no non-Kolmogoroff probability theory. Opposed to this happy situation in which a person has no problems because he ignores problems, constructivism tries to solve fundamental problems. Constructivism tries to understand and to explain why mathematicians the world over accept the Peano axioms for arithmetic and the Kolmogoroff axioms for probability theory. In other cases constructivism tries to find out which axioms, if any, should be accepted for set theory and for geometry.

The last question, which concerns geometry, would lead us far from mathematics, and for that reason I do not want to take it up here. I want only to point out that the whole problem of foundations for mathematics changes radically if we come to doubt that empirical physics is able to decide between Euclidean and non-Euclidean geometry. I should like you to imagine a mathematics without geometry. In such a mathematics we would be concerned only with numbers, sets, and probabilities. We would have to develop theories concerning these entities without being able to take refuge in the paradigm of geometry.

What would that look like? Whereas the axiomaticists (e.g., the Bourbakians) ignore all attempts to construct foundations for mathematical theories (or even despise such attempts as prescientific), constructivists try to explain and justify systems axioms like those of Peano and Kolmogoroff. The constructive method speaks on behalf of cooperation between constructions and axiom systems.

This cooperation is well known in the case of naive arithmetic and analysis. In naive arithmetic and analysis "axiomatic" means that we define certain structures, like groups, combinations, compact spaces, and mass fields.

These structures are defined using systems of propositional forms called axioms. The definitions of the structures are explained and justified by demonstrating that there are important models that satisfy the axioms. Priority is given to the models, which are drawn from naive arithmetic or analysis. Only when these models are important are the axioms accepted.

The language used in these axioms can be restricted to pure logic as, for example, in the cases of group theory and combinatorial theory. But even a structure like Archimedean ordered groups or the structure of topological space uses arithmetic or set-theoretic vocabulary. There is no problem with this, because in the following the axiomatic method is applied only within mathematics.

Problems first arise when we turn to foundations, that is, when we no longer take it to be self-evident that entities like the natural and real numbers or entities like sets and functions are somehow *given*, that these entities stand somehow at our disposal—as if these entities were like flowers in a garden that we have only to give names and then we begin to find out the truth about them.

The foundational problem in arithmetic is not a serious one. No mathematician seriously denies that it is easy to count, for example, with primitive signs like

$$|, ||, |||, \ldots$$

Anyone can even understand the rule of construction for such signs:

$$\Rightarrow | \\ n \Rightarrow n|$$

To this rule we can add further rules for the construction of pairs of signs, for example,

$$\Rightarrow |, n|$$
$$m, n \Rightarrow m|, n|$$

(the pairs that can be constructed are those for which $m < n$ is valid), and also for the construction of triplets such as for addition and multiplication. Using such rules we arrive directly at certain true assertions, for example,

$$| < n|$$
$$m < n \rightarrow m| < n|$$

The second proposition presupposes the logic of implication, $\rightarrow$. No one seriously denies even $\neg n < |$, if one accepts negation, $\neg$, in general. Please note that this is not a matter of axiom choice. For example, if you were to suggest the axiom $\neg n < \|$, you would be immediately contradicted, because $| < \|$ is true. Similarly, $m < n \rightarrow m| < n$ would be ridiculous, because $| < \|$ is true, but $\| < \|$ is not true.

We do not need formal logic in order to be able to make such assertions, but we must know how to argue using propositions (here these propositions have the form of $m < n$) that are constructed with logical operators like $\rightarrow$ and $\neg$. To know how to argue means that we must be acquainted with the dialogical employment of the logical operators. I hope you are familiar with these, for example:

$$A \quad ? \quad \left\| \begin{array}{c} A \rightarrow B \\ B \end{array} \right.$$

The controversy between classical and intuitionistic logicians proves to be a fight over words—for example, whether

$$\neg A \rightarrow \neg B \dot{\rightarrow} B \rightarrow A$$

is logically true (or whether only the converse is logically true). In each case we have to explain and justify our choice of a *general* rule for the dialogue. For both intuitionistic logic and classical logic we can do this by supplying the corresponding Gentzen cut principle. Classical logic proves to be a convenient simplification of intuitionistic logic. As regards junctors, we need only let go all disjunction and implication. (We can then reintroduce them using definitions: $A \vee B$ by $\neg \neg A \wedge \neg B, A \rightarrow B$ by $A \vee B$.) In this way the Peano axioms turn out to be true propositions in constructive arithmetic. ω-incompleteness (first proven by Goedel) demonstrates that not all constructively true propositions are logically deducible from the axioms. This should come as no surprise. A universal proposition $\bigwedge_x A(x)$ is constructively true when $A(n)$ for all n is true. But in order *logically* to deduce the universal proposition $\bigwedge_x A(x)$, we must first deduce $A(x)$ with a free variable x. So we should have expected ω-incompleteness. But Peano arithmetic is ω-complete, if we restrict ourselves to addition. The point of Goedel's proof was to demonstrate that Peano arithmetic with only addition and multiplication (without the higher forms of inductive definition) already shows the ω-incompleteness that was to be expected in general. It is well known, however, that Goedel's theorem of incompleteness of 1931 was a great surprise. To this day it is not widely known that it is necessary that quantification over the elements of the model use only formal logical inference (precisely) because we want to prove propositions for all models. In arithmetic, however, we can obtain from the Peano axioms *all* true arithmetic propositions only if for quantifications over the "natural" numbers we use quantifications over $|, ||, |||\ldots$.

The Peano axioms together with *formal* logic still provide an insufficient foundation for arithmetic. When this is made clear, the fight between axiomatic and constructive methods in arithmetic still remains only a verbal dispute. Because the Peano axioms are constructively true propositions, it is senseless to maintain that mathematics begins only when we have written down some propositions as axioms.

The dispute becomes serious when we come to analysis. From Cauchy to Weierstrass mathematicians in the last century succeeded in defining the real numbers and in proving certain fundamental theorems concerning these numbers, particularly the completeness theorem. The definitions and proofs are no longer arithmetic, because they make use of sets and certain theorems concerning these sets. Sets had not yet been defined in the previous century, and their fundamental theorems—for example, the comprehension principle, $\bigvee_s \bigwedge_x. x \, \varepsilon \, S \leftrightarrow C(x)$, for all propositions of the form $C(x)$—were not yet proven. But it was already clear to Frege that sets should be defined as two propositional forms $A(x)$ and $B(x)$, for which $\bigwedge_x A(x) \leftrightarrow B(x)$ stand for the same *sets*.

If we use $\{z|A(z)\}$ and $\{z|B(z)\}$ for the respective sets, then we have the following definition of identity:

$$\{z|A)z)\} = \{z|B(z)\} \Leftrightarrow \bigwedge_x A(x) \leftrightarrow B(x)$$

For every expression $A(x)$, in which "$A(x)$" occurs such that truth remains invariant, if "$A(x)$" is replaced by a "$B(x)$" that stands for the same set, then we write $\{x|A(x)\}$ instead of $A(x)$. In this way sets are "defined" *by abstraction,* as we say. This is a *façon de parler* for dealing with propositional forms. If $\bigwedge_x A(x) \leftrightarrow B(x)$ is valid and one can abstract from the distinctions between $A(x)$ and $B(x)$ [i.e., $A(x)$ and $B(x)$ occur invariantly], then we replace the propositional forms with the "abstract" object $\{x|A(x)\} = \{x|B(x)\}$. Abstract objects are thus produced by this façon de parler. That is the purpose of definitions by abstraction. After we have abstraction we can define the elementary relation by

$$x\{z|C(z)\} \Leftrightarrow C(x)$$

The comprehension principle follows immediately from this, but with a noteworthy restriction. The formulas A, B, and C may not contain any quantifiers that quantify sets. Otherwise, we would not have a *definition* of sets. Therefore, this constructive basis does not provide a demonstration of the truth of the unrestricted comprehension principle, that is, of the comprehension principle without the restriction of the propositional form $C(x)$ to formulas *without* quantification over sets. We call unrestricted comprehension *nonpredicative*.

All of the modern systems of axioms for set theory are nonpredicative; that is, they contain axioms that imply at least some unrestricted cases of the comprehension principle. Therefore, we have no proof that the axioms of set theory are constructively true. That is the decisive difference between set theory and arithmetic.

Hilbert's attempt to add a logical calculus to the nonpredicative axiom systems and then to prove that the *formal system* that results is at least formally consistent (that for no formula A both A and $\neg A$ are deducible in the formal system) remains for us to complete.

Unhappily, a constructive proof of formal consistency appears to be very difficult. Nonetheless, this difficulty does not mean that no mathematician may engage in analysis so long as the problem of freedom from contradiction is unresolved. There is indeed always the possibility that, having deduced a formula A, the formula $\neg A$ is also deducible. This possibility can be avoided, however, if we perform our analysis constructively instead of trying to formalize the confusion in Cantor's naive set theory.

If you have proven a proposition A constructively, then you know that $\neg A$ is not constructively provable.

Constructive analysis means that we begin with constructive arithmetic, proceed to rational numbers, and then construct the propositional forms $A(r)$, $B(r)$ with a free variable r for rational numbers. The propositional forms are taken only from arithmetic—that is, using inductive definition and combination using logical operators, with quantifiers operating only over natural or rational numbers. If we then use only restricted comprehension, we are led to the definition of sets of rational numbers and of real numbers. Real numbers are here again defined by abstraction.

The identity $\text{fin}_r A(r) = \text{fin}_r B(r)$ is defined as follows: The sets $\{r|A(r)\}$ and $\{r|B(r)\}$ have the same *classes* of rational numbers. The restriction that we can have no set quantifiers in the propositional forms implies that we also have no quantifiers over real numbers in the comprehension principle. The point of constructive analysis, as I have shown in my book on differentials and integrals, consists in the fact that analysis can be conducted essentially as was done previously despite this restriction. For all the applications of analysis (excepting pure mathematics), the constructive method of analysis proves to be adequate.

Analysis requires much greater changes if classical logic is replaced by intuitionistic logic or if functions are to be restricted to recursive functions. For that reason, constructive analysis is often said to be complicated and inconvenient. But that is at best true of versions involving intuitionism or recursive analysis. Constructive analysis in my sense, which places a restriction only on the principle of comprehension, has the same elegance as analysis based on axiomatic set theory. The only difference is that the constructive version avoids the danger of contradiction from the outset. Therefore, it also needs no proof of freedom from contradiction.

The advantage of constructive analysis in all its forms is that we are freed of the pluralism of axiomatic systems. If we were to have a proof of freedom of contradiction for ZF (Zermelo–Fraenkel system), the work of Goedel and Cohen would at the same time provide a basis for proofs of freedom from contradiction for ZF $\wedge$ CH as well as for ZF $\wedge$ ⌐CH. There would then be no rational basis for deciding among these formal systems. We would continue to be faced with the present plurality of irrational decisions within which only the indifference that we call tolerance produces that illusion of a common endeavor. The differences among the various constructive investigations are not pluralism; rather, they are indicative only of different emphases. Constructivism in the nonrecursive sense of my book includes recursive constructivism, because the restriction to recursive functions only marks out certain functions as particularly important. On the other hand, intuitionistic logic is a logic that includes classical logic within it, as I have already indicated. Therefore, the restriction to classical logic is not an unreasonable choice but rather, for the sake of convenience, is one that confines itself to so-called classical existence, $\neg \bigwedge_x \neg A(x)$, instead of constructive existence, $\bigvee_x A(x)$. An intuitionistic nonrecursive constructivism would be an all-encompassing theory, with the other constructive theories being special theories under it.

In contrast to this peaceful coexistence, we have a pluralism of axiomatic theories that contradict one another as do CH and ⌐CH. Since the advent of non-Euclidean geometry, mathematicians have grown comfortable with this contradictory pluralism. But that is a problem that we do not want to go into here. I would much rather conclude with an attempt to make intelligible the noteworthy fact that in this pluralistic world there is still only one axiomatic theory of probability; there is no non-Kolmogoroff theory of probability.

It seems simple enough to define probability. Within analysis—it does not matter here whether it is constructive or axiomatic analysis—we can, for example, define Kolmogoroff fields as σ-fields of sets with a normalized σ-additive measure. That is no problem. The problem is why these Kolmogoroff fields are called probability fields and how the term "probability" can be defined so that we can justify application of these fields in all areas of modern statistics. This problem of justifying applicability is the problem referred to when we speak of justifying the Kolmogoroff axioms.

Let me begin with descriptive statistics, where a set, called a *population*, of N elements $C_1, \ldots C_N$ is given. We may know that n elements of the population satisfy a certain propositional form $A(x)$. The frequency of A (usually called the *relative frequency*) is defined by

$$\rho(A) = \frac{n}{N}.$$

Using elementary arithmetic, there immediately results

(1) $\rho(A) = 1$, if all elements satisfy A
(2) $\rho(A \vee B) = \rho(A) + \rho(B)$, if A and B are disjoint

Now suppose that an element of the population is chosen. We do not know which. We will, however, give to this element the name "c." Now we have to make predictions concerning $A(c)$. Predictions are formulated using modalities. If $\rho(A) = 1$, then $A(c)$ would be *necessary*. If $\rho(A) = 0$, then $A(c)$ would be *impossible*. For $0 < \rho(A) < 1$, we claim that $A(c)$ is *contingent* (i.e., neither necessary nor impossible).

Next it is proposed that we make use of a comparative language and that it be said that the more $\rho(A) = 1$, the more necessary $A(c)$ is. Then it is proposed that this comparative language be tightened into a quantitative language and that it be said that $A(c)$ has the "degree of necessity" $\rho(A)$. In place of "degree of necessity" we can then also say "probability."

This last proposal, however, makes no sense in general. For example, consider the population of a city in which the frequency of women and men is one-half. Now, were we to take the case of the first person that we see in the morning, the probability that this person would be a woman would be one-half. Now, at least in the case of a faithful husband, this is absurd, because he sees his wife first every day.

This example demonstrates that the element c must be chosen *randomly*. That is, in order to justify the introduction of probabilities we must first *define* chance (randomness). Indeed, we must define chance without presupposing probabilities. This can be done with the help of gaming apparatuses like, for example, dice or roulette wheels. We shall call such apparatuses random generators. In a specific instance, we require that a random generator be an apparatus that satisfies the following conditions:

1. *Uniqueness:* Each use of the apparatus produces as a result exactly one of finitely many propositional forms $E_1, \ldots, E_m$ (elementary events).
2. *Repeatability:* Following each use, the apparatus is in the same condition as prior to that use.
3. *Undecidability:* There is no causal knowledge on the basis of which the result could be determined *prior* to a specific use of the apparatus.

This definition of a random generator does not belong to mathematics. It is a prescription for making such apparatuses. In the following I will assume that engineers can understand this language and that our technology is able to produce such random generators. No actual random generator will be *perfect,* but the actual ones are good enough to justify the concept of a random generator as it is defined here by uniqueness, repeatability, and undecidability.

As regards the foundation of probability theory, we proceed on the basis of a historical fact—the fact that in our culture we are able to produce generators that are sufficiently random. The "empiricists" among philosophers or the theoreticians of science will call this fact an experience. But it should be noted that, corresponding to the production of sufficiently flat surfaces in geometry, it is here a matter of prescientific experience, of so-called life experiences, which are prior to all experiences that are to be assessed scientifically. In the very articulation of this fact the expression "sufficiently . . ." places an obstacle in the way of a scientific assessment. Nonetheless, the production techniques are well enough developed that it is reasonable to take the ideal norms as the basis for scientific theories.

The results produced by a random generator are contingent. We only propose to define a probability p for the propositional forms $E\mu_1 \vee \ldots \vee E\mu_r$ (i.e., the disjunction of elementary events) such that the following three conditions are satisfied:

(1) $p(A) = 1$, if A is necessary
(2) $p(A \vee B) = p(A) + p(B)$, if A and B are mutually exclusive;
(3) $p(E_1) = \ldots = p(E_m)$

We require that p satisfy conditions 1 and 2 for frequencies and we add condition 3 because of the requirement that random generators be undecidable. Conditions 1–3 clearly and immediately imply

$$p(E_{\mu_1} \vee \ldots \vee E_{\mu_r}) = \frac{r}{m} \text{ for } \mu_i \neq \mu_j$$

which is to say that we have a Laplace field of probabilities. The definition of random generator certainly justifies 3. But why are we justified in requiring that the "degree of necessity" satisfy conditions 1 and 2, which are valid for frequencies? The reason lies in Bernoulli's law of large numbers. If we use a random generator in a series of trials of length L, then on the one hand we must calculate $p(A)$ according to 1–3, and on the other hand we can observe the frequency $\rho L(A)$. There are m^L results of such a series of trials. Due to repeatability, all these results are undecidable. Therefore, we can calculate a probability p^L for all their disjunctions, in particular for the propositional form $\rho_L(A) - p(A) | < \varepsilon$ for every positive ε.

Bernoulli's law says

$$\lim_{L \to \infty} p_L[|\rho_L(A) - p(A)| < \varepsilon] = 1$$

which can be roughly expressed as meaning that the probability $p(A)$ is simply the frequency $\rho_L(A)$ "in the long run." The pure mathematical proof of Bernoulli's law justifies defining probabilities with the help of conditions 1 and 2, which are true for frequencies.

In the case of continuous random generators, there are no elementary events. A roulette wheel is arbitrarily divided into a finite number of intervals of equal length. If the circumference has a length of 1, then the probability of each interval is precisely its length. Similarly, for a two-dimensional apparatus the areas of rectangles represent the apparatus's probabilities. Because it would be arbitrary to restrict this definition of probability to rectangles, probability is extended to Lebesque measurable sets. This means that condition 2 is strengthened to σ-additivity for the series $A_1, A_2, A_3, \ldots$

(2σ) $\qquad\qquad p(V_r A_r) = \Sigma_r p(Ar)$

if all A_{r1}, A_{r2} are mutually exclusive.

Constructive and Axiomatic Methods | 247

In this way we can move from continuous random generators to Lebesque fields of probabilities in n-dimensional number space.

Once we are technologically able to produce random generators with Laplace or Lebesque probability fields, it is easy to produce further probability fields by combining random generators into *random aggregates*.

If the generators are causally independent (a task for engineers), we can then define product fields by using several generators at once. This product will also be a model for the Kolmogoroff axioms 1 and 2. Further operations using random aggregates produce additional probability fields:

1. *Relativization*, which means that we define

$$p_B(A) = \frac{p(A \wedge B)}{p(B)}$$

   for every B with $p(B) > 0$

2. *Coarsening*, in which a σ-field is homomorphically projected onto a new σ-field.

For example, in the continous case coarsening is the result of various types of loaded dice. Suppose we have a square quadrilateral with its center of gravity S lying outside of its geometrical center (fig. 20.1). If the probability field for impact in an interval of the circumference (which we mentally circumscribe) is a Lebesque field, then the probability field for

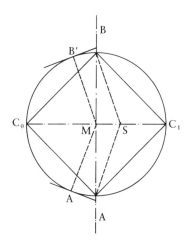

FIGURE 20.1

the four sides is a discrete coarsening but not a Laplace field. Clearly, we can similarly obtain all discrete fields by coarsening Lebesque fields. Finally, Kolmogoroff showed that the outcomes of stochastic processes (which are determined at every point only by a probability field) represent the events of a probability field.

Therefore, since Kolmogoroff (1933), mathematical probability theory has been nothing other than the theory of σ-fields with a normalized σ-additive measure.

By trivially asserting that we have thereby defined what "mathematical probability theory" is, mathematicians evade further asking whether they are really interested in *all* models of the axiom system $1-2\ \sigma$.

The Kolmogoroff statement concerning the outcomes of stochastic processes is provable only under the additional condition that it concerns the Borel fields of certain topological spaces, specifically, so-called polish spaces. In fact, all probability fields that can be defined using random aggregates satisfy these additional topological "axioms."

Moreover, every "polish" Kolmogoroff field is representable as the limit (in the sense of so-called indeterminate convergence) of discrete probability fields. It is only this approximation principle that justifies calling all polish Kolmogoroff fields "probability fields." The remaining models of the Kolmogoroff axioms are irrelevant for probability theory. We cannot evade this irrelevance by nonetheless *calling* all models of Kolmogoroff axioms "probability fields."

In application to natural processes (where causal explanation fails) we try the fiction that the process occurs *as if* it were determined by a (hidden and unknown) random aggregate. We do not try to ascertain the elements of this fictitious aggregate; rather, we only try to find a usable approximation to the probability field. That is the task of statistical theory.

With this last example—a constructive foundation for axiomatic probability theory (in which I have constructed the random aggregate as a model)—I want to end this discussion. The constructive method makes it possible to go beyond the irrational pluralism of merely axiomatic mathematics. The axiomatic method, which is the "conceptually economical" way to handle already constructed models, is one of the greatest achievements of the nineteenth and twentieth centuries. The attempt to use axioms in place of constructive foundations, however, leads only to our implying that an axiomatic "foundation" has been produced wherever we happen to have written down some sentences without foundation. That we call these sentences "axioms" changes nothing. Despite the surpassing importance of the axiomatic method in (naive or constructive) mathematics, the expression "axiomatic foundation" remains nonsense.

TWENTY-ONE

Concerning a Definition of Probability

In statistics, in which probability theory finds its application, little time is normally lost in defining the term "probability." Only in philosophy of science do we find a protracted and continuing controversy concerning the, as the saying goes, concept of probability.

If we ignore the special terminology of philosophy (like "subjective–objective"), we have in essence the "empiricists," who want to relate "probability" back to *observed* (possibly idealized) frequencies, arrayed against the "a priorists," who want to do without observations (i.e., without anything "empirical").

In the following I will propose a definition that, to be sure, makes use of frequencies but also adds to that an "idealization" using a normative concept of chance or randomness (via norms for the construction of random generators). The law of large numbers is then a *proof* that probabilities are idealized frequencies.

Let there be given a set (population) of N elements $c_1, \ldots c_N$ and a description of the set to the effect that n elements satisfy a specific propositional form $A(x)$. The frequency of A—usually called the *relative frequency*—is defined by

$$\rho(A) = \frac{n}{N}$$

According to elementary arithmetic, then, the following propositions are valid for frequencies.

(1) $\rho(A) = 1$,
(2) $\rho(A \vee B) = \rho(A) + \rho(B)$ for disjunct propositional forms A and B

If we define the conditional frequency $\rho_A(B)$ using the frequency of B in the subset of elements with A, then the following is valid:

$$\rho_A(B) = \frac{\rho(A \wedge B)}{\rho(A)}$$

If we take an element c as a "test sample" from the population, then we say that "the probability is $\rho(A)$" that $A(c)$ will be true. It is a matter of predicting $A(c)$. If $\rho(A) = 1$, then $A(c)$ is *necessary*, whereas if $\rho(A) = 0$, then $A(c)$ is *impossible*. In cases of $0 < \rho(A) < 1$, $A(c)$ is *contingent*, that is, possible but not necessary.

Probability theory must establish the conditions under which this contingency assertion can meaningfully be made more precise by adding that the closer $\rho(A)$ approaches 1, the more the contingency becomes a necessity or, to use a comparative, the "more necessary" $A(c)$ becomes.

For quantitative precision we can then introduce $\rho(A)$ as representing the "degree of necessity" of $A(c)$. Rather than "degree of necessity," it is customary to use the expression "probability," which is introduced as the translation of *verisimilitudo*. The "more probable," the greater the similitude to (necessary) truth.

It is not meaningful to determine the probability $\rho(A)$ as a more precise measure of the contingency of $A(c)$ by taking a "test sample," if *just any* selection of an element is going to be called a "test sample." If we *deliberately* take a c with $A(c)$, then $A(c)$ would be necessary even for $\rho(A) < 1$. In taking a "test sample" an element must be *randomly* selected.

To provide a foundation for probability propositions we must therefore next define "chance" or "randomness." We do this by making use of *random generators* like dice or roulette wheels. An apparatus is called a "random generator" if it satisfies the following conditions:

1. *Uniqueness:* Each use of the apparatus (each trial) produces a result that is exactly one of finitely many propositional forms $E_1, \ldots, E_m$ (elementary events).
2. *Undecidability:* There is no causal knowledge that makes it possible to distinguish beforehand one of the results $E_1, \ldots, E_m$ from the others.
3. *Repeatability:* After each trial the apparatus is in the same condition as prior to the trial.

To this foundation of probability theory we will add the historical fact that our culture is able to produce technically "good" random generators. There are no "perfect" random generators, but we do have sufficiently good realizations of the norms of uniqueness, undecidability, and repeatability. We need to establish that, if we "randomly" selected (i.e., with the help of a random generator) an element c from a set $c_1, \ldots, c_N$ with N elements and if $\rho(A)$ is the frequency of the propositional form A, the probability $\rho(A)$ is given to the prediction $A(c)$. A definition alone is not sufficient, because we could certainly define $\rho(A)$ as the probability of $A(c)$ also in cases of nonrandom selection, but this—as we have seen— would be unreasonable.

We obtain a foundation in the case of *random* selection of c, if we next assign a probability ω for every c_ν of the proposition $c = c_\nu$.

Because $c = c_1 \vee \ldots \vee c = c_N$ is necessary due to the uniqueness of the random generator, in accordance with (1) we can posit

$$\omega(c = c_1 \vee \ldots \vee c = c_N) = 1$$

Because the propositions $c = c_1, \ldots, c = c_N$ are also pairwise mutually exclusive due to the uniqueness of the random generator, in accordance with (2) we can posit

$$\omega(c = c_1) + \omega(c = c_2) + \ldots + \omega(c = c_N) = 1$$

Finally, on the basis of the undecidability of the random generator we can posit

$$\omega(c = c_1) = \omega(c = c_2) = \ldots = \omega(c = c_N)$$

For all $\nu = 1, \ldots, N$, this yields

$$\omega(c = c_\nu) = \frac{1}{N}$$

If A has the frequency $\rho(A)$ in $c_1, \ldots, c_N$, there are $\rho(A)$ times N elements c_ν with $A(c_\nu)$. The probability for c's being one of these c_ν is therefore

$$\rho(A) \cdot N \cdot \frac{1}{N}, \text{ i.e.,}$$

$$\omega[A(c)] = \rho(A)$$

according to (2). To obtain this equation, we have postulated the following for propositions about the results $<$ of a random generator in order to calculate a probability ω:

(1) $\omega(A) = 1$, if $A(c)$ is necessary
(2) $\omega(A \vee B) = \omega(A) + \omega(B)$, if $A(c)$ and $B(c)$ are mutually exclusive
(3) $\omega(E_1) = \omega(E_2) = \ldots = \omega(E_m)$

For the sake of simplicity we have—as is customary—written $\omega(A)$ instead of $\omega[A(c)]$.

Conditions 1 and 2 rest upon the corresponding principles concerning frequencies, whereas condition 3 rests upon the undecidability of random generators.

Before carrying out a trial using a random generator, every elementary proposition concerning the result $E_\mu(c)$ for $\mu = 1, \ldots, m$ is contingent. But because of undecidability we are dealing with a special contingency. Its specialness lies in the fact that here—as in the case of roulette—a wealth of causal knowledge has been used in order to so *construct* these machines that (to the best of our causal knowledge) no distinction between the possible results can be discerned before they occur. We cannot say on the basis of the construction of a roulette wheel that "red" is "more necessary" than "black." The requirement of undecidability is a requirement placed upon the construction of random generators. We know more about $E_\mu(c)$ than simply that it is contingent. I suggest that the propositions $E_\mu(c)$ be called *fully contingent* or *random,* for short. In this terminology a random "generator" is an apparatus (experience = lab design) that produces *random* results.

For the combination 1–3 of the calculation of frequencies with random generators we must still clarify to what extent a frequency can be assigned to the results of a random generator.

Using a random generator, let there be conducted a series of trials of length L. For every proposition A consisting of disjunctive combinations from $E_1, \ldots, E_m$, first calculate the probability $\omega(A)$ using 1–3, and then calculate the frequency $\rho_L(A)$ of A in the series of trials of length L. For every positive ε also calculate a probability ω_L of $|\rho_L(A) - \omega(A)| < \varepsilon$. The proposition $|\rho_L(A) - \omega(A)| < \varepsilon$ is a proposition concerning *arbitrary* series of trials of length L. There are m^L such series of trials. $\omega_L(|\rho_L(A) - \omega(A)| < \varepsilon)$ is the frequency of series of trials of length L that satisfy the proposition $|\rho_L(A) - \omega(A)| < \varepsilon$. (This frequency is the desired probability, because all m^L series of trials—more precisely, the m^L results of the series of trials of length L—have the same probability. Due to repeatability, these m^L results are also undecidable.) Now, according to Bernoulli, the probability of $\omega_L(|\rho_L(A) - \omega(A)| < \varepsilon)$ converges on 1 for $L \to \infty$. That is the "weak law of large numbers:"

$$\lim_{L \to \infty} \omega_L(|\rho_L(A) - \omega(A)| < \varepsilon) = 1$$

That means: With increasing L the proposition that the frequency $\rho_L(A)$ up to ε is exactly $\omega(A)$ comes arbitrarily close to a necessity. Expressed less precisely, probability *is* the "frequency over the long run." It is only this Bernoullian law for propositions about trials using random generators that justifies defining "probability" as a numerical frequency calculated according to 1–3.

In the case of continuous random generators (e.g., roulette wheels), there are no elementary events. A roulette wheel does in fact stop at a specific spot, but it would make no sense to assign a (positive) probability to every "point." Instead, we divide the circumference arbitrarily into a finite number of intervals of *equal length*—and each of these intervals possesses the *same* probability. If the circumference has a length of 1, then the probability of an interval of a given length is simply its length. In the case of a random generator that stops in a "random" manner at a spot on a rectangle (the area of which equals 1), we define—under corresponding conditions—the probability that this spot will lie within a given portion within the rectangle to be the area of that portion. This definition is precisely as "arbitrary" or "rational" as the definition of the area of a rectangle as the limit of the sum of the areas of the subrectangles.

We can tighten up condition 2 using "complete additivity"

$$(2\sigma) \quad \omega(V_v A_v) = \Sigma_v \omega(A_v)$$

for every sequence A_*, with A_{v1}, A_{v2} being mutually exclusive. Using the results of the modern theory of measurement (Borel, Lebesque, Frechet), Kolmogoroff (1933) was able to demonstrate that this proposed definition of probability always results, in set-theoretic terms, in defining a normalized, completely additive measure in terms of a σ-field; that is, it always results in a probability field. Discrete random generators produce Laplacean probability fields. Continuous random generators produce Lebesque probability fields in n-dimensional space. If we build on von Mises, it can also be demonstrated that combining random generators into *random aggregates* always produces probability fields that in general are neither Laplacean nor Lebesque.

These combinations correspond to operations that apply to probability fields. As regards *a* probability field, these operations are *relativizations* (*Teilungen*, in von Mises), in which we move from $\omega(A)$ to

$$\omega_B(A) = \frac{\omega(A \wedge B)}{\omega(B)}$$

for a set B with $\omega(B) > 0$, and *coarsening* (*Mischungen*), in which the field is homomorphically transformed into a different field. Any bowl containing balls (according to their colors) can be such a coarsening insofar as the balls (and their evidentiary marks) can be grouped into sets. A "loaded die" produces a probability field that results from the coarsening of a Lebesque field. We can reduce the problem to two dimensions by "throwing" a square prism instead of a die (see fig. 21.1). The center of mass S must be distinguished from the geometric center. We imagine that the prism C_0AC_1B is enclosed in a cylinder. The prism is then "thrown" using a random generator so that—were it cylindrical—it would fall straight onto a table without spin and with equal probability of striking the table with equal segments of the (imagined) circumference. Without the cylinder, the prism will fall on one of the four sides. We want to know the probability that C_0 will be on top. If A' or B' is so situated on the circumference that $A'M$ of $B'M$, respectively is parallel to AS or BS, respectively, then the length of the arc $A'C_1B'$ is the desired probability. If the position of S is known, then this probability $\omega > \frac{1}{2}$ can be calculated. Unless we already know the center of mass, we can obtain only the "well-supported" hypothesis $\omega = C^L$ concerning the probability by determining the frequency C^L of C_0 in sufficiently long series of trials of length L according to Bernoulli's theorem; that is, the frequency $C^L \pm \varepsilon$ would have for small ε on this hypothesis a probability ω_L near 1.

More important than relativization and coarsening is the *product* (*Verbindung*) of a number of independent random aggregates. Only by means of this relation do we move from generators to aggregates. Here—as in the case of the repeatability of generators—this independence must be defined as technically realizable causal independence: To the best of our causal knowledge a trial with one aggregate effects no alteration of the other aggregates. Mathematically, the product of probability fields is a new probability field, the product field. Only mathematically is it possible

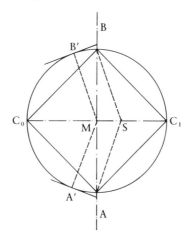

FIGURE 21.1

to multiply (countably) infinitely many fields, which gives—as, for example, in the strong law of large numbers—an agreeable possibility of talking about the products of finitely but arbitrarily many fields.

Kolmogoroff further demonstrated that the course of stochastic *processes* (which at any given point in time are determined only by a probability field) can also be treated as "events" in a probability field.

Since Kolmogoroff, "mathematical probability theory" has therefore been conducted as the theory of σ-fields with a "probability" that axiomatically requires only 1 and 2 σ.

I would like to suggest that we show a bit more caution and call any model of the Kolmogoroff axioms a *Kolmogoroff field*. All models that we obtain when we start with Laplacean and Lebesque fields and then apply all operations (relativization, coarsening, product formation, and process formation) satisfy additional "axioms." In all applications we are dealing with Borel fields of topological space. It has been demonstrated that the classes of so-called polish space (which are separable and completely metric) are large enough to be closed in relation to all proposed operations for probability fields.

The class of "polish" Kolmogoroff fields—in distinction from the class of all Kolmogoroff fields—can on the other hand be proven to be not too large; all of these fields can be represented as the limits of discrete probability fields. The concept of "limit" that is used here has been given precision by H. Cartan. He called it "vague convergence." This approximation theorem supplies the justification for calling all models of normalized σ-additive measures for the Borel fields of polish spaces *probability fields*.

The relevance of the approximation theorem rests naturally on its applications.

In specific applications we must again refer to a *definition* of the concept of probability. This situation is disguised by the fact that in statistical applications (which are much more important than the original applications to gambling) the random aggregates do not occur explicitly but are only hypothetically imputed. Just as the inertial masses of the sun and planets are not "theoretical quantities," probabilities in statistics are also not "theoretical quantities." Rather, in statistics we work with the fiction that it is as if the observations are the results of unknown (concealed) random aggregates, "as if God rolled the dice." The decay of radioactive material, for example—which observation shows can be described as exponential decay—is "explained" as the frequency of decaying atoms (if x is the number of atoms and dx—formulated as an infinitesimal—the number of decaying atoms in the time interval dt) is proportional to dt:

$$\frac{dx}{x} \sim dt$$

Using integration, it follows from this that $\Delta \lg x = - K\Delta t$ for a positive constant K, that is, the exponential law:

$$x = x_0 e^{-K(t-t_0)}$$

The frequency of decay $K\Delta t$ in a (sufficiently small) time interval Δt is for its part "explained" as each atom's having a *probability* of decaying in Δt of $K\Delta t$. It is thus asserted that the decay of atoms occurs "as if" a random aggregate "removed" the atoms.

How statistics (in physics or any other science) solves individually the problems of such statistical hypotheses (particularly the problem of their verification) is not a part of philosophy of science, which has only the problem of finding a definition for the fundamental concept of "probability" and providing a justification of the concept's applicability.

TWENTY-TWO

The Foundational Problem of Geometry

If we take the word "science" in a sufficiently restricted sense so that not every body of practical knowledge falls under it, then geometry is the oldest science. In contrast, the problem of a foundation for geometry as an independent problem is, remarkably, only about 100 years old. It has only been since the so-called discovery of non-Euclidean geometry that anyone has seriously been concerned with whether and how we could ascertain and justify which geometry is the *true* geometry.

To explain this situation and as preparation for a solution to the foundation problem, I will briefly recount the history of geometry. We can very roughly distinguish three periods in this history: (1) the geometry of the ancient Middle East (Mesopotamia and Egypt), (2) the geometry of the Greeks, and (3) modern geometry.

Ancient Middle Eastern geometry was by no means primitive. For example, around 2000 B.C. it was known in Babylon how to apply what we call the Pythagorean formula with the same precision that we achieve. We find in cuneiform texts, for example, the following: A beam that is five units long has slid out from a perpendicular wall, losing one unit of height (fig. 22.1). How far out from the wall does the beam extend?

Without further commentary the text next calculates the one side of

FIGURE 22.1

the resulting triangle to be $5 - 1 = 4$ units and then calculates the requested side to be $\sqrt{(5^2 - 4^2)} = 3$ units.

We should not, however, inquire after a proof of the validity of this reckoning procedure. The concept of a proof did not exist in ancient Middle Eastern geometry. Rather, the Greeks' extraordinary productive discovery of geometry as a science consisted precisely in their having seen and realized the possibility of elaborating practical geometrical skill into a *system* of propositions that are mutually bound together into proofs by chains of logical inferences. Today we still say, using the Aristotelian term, that geometry thereby becomes an *axiomatic theory*. Certain propositions are placed at the beginning of the theory without any proof; these are the axioms. All other propositions are then to be logically deduced from the axioms. In this, as you can see, geometry presupposes logic as a science, or at least the ability to draw logical inferences.

According to the Greek conception, geometry is also a general theory of quantity that contains arithmetic as a special instance, namely, as the theory of discrete quantities. Nonetheless, the relation of geometry to arithmetic is of a different sort. In addition to logic, Greek geometry also presupposed arithmetic, for example, in the case of the so-called Archimedean axiom. This axiom is found in Euclid and can be traced with confidence back to Eudoxus. The axiom says that the repeated halving of a length always results after a finite number of steps in a point in each arbitrarily chosen partial length. The "after a finite number of steps" that appears here signifies the existence of a natural number. The metamathematical investigations of Tarski have succeeded in proving that the Archimedean axiom cannot be replaced by any combination of non-arithmetical propositions.

The relation of logic, arithmetic, and geometry can be restated in the following construction metaphor. The edifice of the sciences is raised upon a foundation of prescientific practical skill. In figure 22.2 I have designated a lower, nonlogical part of arithmetic upon which arithmetic is

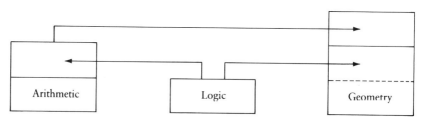

FIGURE 22.2

elaborated with the help of logic. In contrast, the current ruling opinion holds arithmetic to be a science only insofar as it is an axiomatic theory. This denies the legitimacy of a constructive arithmetic. I want here, however, to leave this dispute to the side.

According to the Greeks, there is no nonlogical foundation for geometry. In this connection we should note Plato, who ridiculed geometers for beginning with certain propositions as if neither they nor anyone else were accountable for these propositions.

Modern geometry is sharply distinguished from Greek geometry. This distinction results, above all, from three novelties:

1. Analytic geometry (Descartes, 1637)
2. Non-Euclidean geometries (Lobachevski, 1829; Reimann, 1854)
3. Gravitation geometry (Einstein, 1915).

I have left projective geometry out of this enumeration, because in my judgment it plays no role in the foundation problem—no more than does Klein's Erlanger Program.

The appearance of analytic geometry produced for the first time a translation of all geometric propositions into arithmetic propositions. Since that time arithmetic has supplied the model for the geometrical axioms, here called the Cartesian model. In the Cartesian model the geometric axioms are provable propositions. Guided by its extension through the infinitesimal calculus, arithmetic during the eighteenth century was so dominant that the method of axiomatic theories was pushed to the boundaries of mathematics. Only the weight of tradition standing behind the Euclidean texts accounts for the fact that, even at the end, people continued to further investigate the axiomatic basis of Euclidean geometry.

The nineteenth century succeeded in solving a problem discussed since antiquity; analytic geometry made it possible to demonstrate that the so-called parallel axiom (according to which in every plane for a given line and a given point there is at most one line through the point that does not intersect the first line) *cannot* be logically deduced from the other axioms alone. This demonstration of the logical independence of the parallel axiom was the so-called discovery of non-Euclidean geometries. The first publication concerning these geometries was that of Lobachevski in 1829. Gauss had in fact obtained this result earlier but had not published it, because in his investigations he was concerned not only with this logical independence but also with a good deal more.

Gauss began with the position that distance, the interval between two points, could be taken as the sole fundamental concept of geometry.

Every curved surface then supplies an example of a non-Euclidean two-dimensional distance geometry. Opposing Kant, Gauss held it to be a meaningful empirical question to ask whether real space is curved or flat. In his inaugural dissertation of 1854 Reimann spoke of the hypotheses that underlie geometry. If the curvature of space is an empirical question, then physics must answer this question. Helmholtz attacked the question in 1868 in his investigations concerning the facts that underlie geometry. He then proposed a mathematical theory, the goal of which was to demonstrate which of the Euclidean axioms were valid in a distance geometry, that is, a geometry in which the concept of distance satisfied only those requirements that are required to call it a distance geometry at all. Helmholtz's theory was extended by Lie and in recent years by Busemann. This has now led to the result that under a few self-evident requirements on the concept of distance all of Euclid's axioms are satisfied with the sole exception of the parallel axiom.

Most important among the self-evident requirements posited by this distance concept is unrestricted mobility. This requirement says that every tetrahedron can be transformed into any other tetrahedron with sides of the same length through a distance-preserving mapping of the whole space. The investigations of Helmholtz, Lie, and Busemann have given a foundation to absolute geometry in the form of distance geometry—absolute geometry is the name given to all that can be proven in Euclidean geometry without the parallel axiom.

Of course, this foundation is only a matter of a mathematical theory. This theory demonstrates that the axioms of absolute geometry can be deduced from the distance axioms. Therefore, Helmholtz added a physical foundation to the foundation provided by the distance axioms. He investigated the procedures with which we measure distance. The measurement of distance can be reduced to counting; you count how many times a measuring rod or ruler can be laid down between two points. Conceptually, this laying down means that you use the congruence of two pairs of points as a fundamental concept. If a measuring rod is moved from one location to another, then the before and after end points represent congruent pairs of points. Thus, all the concepts of geometry can be physically defined, if you can physically define what a measuring rod is. Now, any rigid body on which two points have been marked can serve as a measuring rod. The problem of a foundation for geometry thus reduces to the problem of defining rigid bodies.

As long as we have available only Euclidean geometry construcd as an axiomatic theory, we neither need nor are able to pose these questions. Now, after it has been demonstrated that the concept of distance (more precisely, the distance axioms) leads only to absolute geometry, there remains the problem of whether and how we might choose between Euclidean and non-Euclidean geometries as regards physically rigid bodies. It appears to be a simple matter. You produce a triangle out of a rigid body and then extend two sides to double their original length (fig. 22.3). Is the

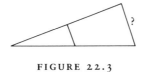

FIGURE 22.3

third side now double in length? It will be precisely double, if the geometry of rigid bodies is Euclidean.

Of course, everything depends on the definition of rigidity. You might say that "rigid" should mean maximally unalterable. But that is only an explication of a word. As Lotze had previously indicated in his metaphysics, Poincaré observed that unalterability must be judged using physical laws. We could also change physics—and leave geometry Euclidean. The discussion of this issue acquired a sense of urgency only with the appearance of Einsteinian gravitation theory. This theory rested upon electrodynamics and convincingly demonstrated that optically measured distances must appeal to a non-Euclidean geometry, if gravitation and electrodynamics were to be united in a physical theory.

So since Einstein we have had a physical geometry. Most physicists believe that the problem of a foundation for geometry has been solved; it is a physical problem, which to this point in time has been best solved by Einstein's theory. In any event, this removes the problem from the purview of philosophy.

This introduction has carried us far from philosophy. However, because the one sure sign of philosophy is untiring critical reflection upon what is thought and supposedly known, up to now philosophy has not tired of asking whether mathematical geometry on the one hand and modern physical geometry on the other hand really exhaust the possibilities of geometry as the science of space, that is, as the science of spatial ordering. Leibniz once offered the definition" "Spatium est ordo coexistendi."

It is true that our rulers are massive and that they consist of electrically charged particles—whence modern physical geometry draws its justification—but it still remains for us to inquire whether there is a geometry that need take no notice of physics and yet is not thereby nothing more than a merely formal mathematical theory.

This question naturally makes one think of Kant and of synthetic and a priori geometric judgments founded upon pure intuition. This appeal to Kant is in vain, however. Nowhere in Kant can we find an explanation of how the axioms of Euclidean geometry (in Hilbert's system there are twenty-seven) can be founded step by step on pure intuition. If only there were somewhere in Kant something that gave a sufficient indication of how to proceed, then all of post-Kantian positivism could not have explicitly denied the existence of a priori and synthetic judgments.

The beginnings of a geometry that is neither merely analytic—that is, a formal axiomatic theory—nor yet an a posteriori theory can be found in the work of Dingler. With this we arrive at the solution to the foundation problem in geometry that I want to propose here. To be sure, in its execution my presentation will be distinct from that of Dingler. The effect of this will be, I hope, that many misunderstandings can be disposed of— misunderstandings that did and do beset Dingler's geometry.

I want to begin with the observation that within Euclidean geometry as an axiomatic theory we can indeed reduce the fundamental concepts—as represented—to congruence but that we can also completely dispense with congruence.

To begin with, we take (as did Euclid and even Hilbert) planes $E, \ldots$, lines $l, \ldots$, and points $P, \ldots$ as fundamental objects. As fundamental relations between these objects let us take besides equality also the so-called topological relation of incidence ($P \, \varepsilon \, E$ or $P \, \varepsilon \, l$, where P lies upon E or l, respectively), and the betweenness relation B [$B(P_1 E P_2)$, abbreviated $P_1 E P_2$, where E lies between P_1 and P_2], both of which relations hold under deformation and therefore are immaterial for defining congruence. Both of the following concepts are decisive:

$$\text{parallelism } \|: E_1 \| E_2$$
$$\text{orthogonality } \dashv: l \dashv E$$

The choice of these fundamental concepts at first appears arbitrary. Nonetheless, these concepts suffice to define congruence. As our first step we can define the parallel congruence of point-pairs P_1, Q_1 and P_2, Q_2 via the existence of a parallelogram with these points as corners (see fig.

FIGURE 22.4

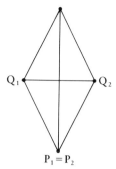

FIGURE 22.5

22.4). Clearly, we need only our fundamental concepts for this. In addition, mirror congruence can be defined—most easily for the special case of $P_1 = P_2$ (fig. 22.5). Only when the diagnosis of the parallelogram are orthogonal (it is then called a rhombus) are P_1, Q_1 and P_2, Q_2 called mirror-congruent.

Finally, in general two pairs of points are called congruent if there is a third pair of points that is either parallel or mirror-congruent to each of the first two pairs (fig. 22.6). In the case of three or more dimensions, the parallelogram and rhombus need not lie in the same plane. When we have defined congruence in this way, a body is said to be geometrically rigid if each pair of points marked on it is always transformed into a congruent pair of points by any movement of the body. Applied to several bodies, we thereby obtain a criterion for rigidity.

In this way the foundation problem is reducible to our new fundamental concepts. Must these concepts now function as merely undefin-

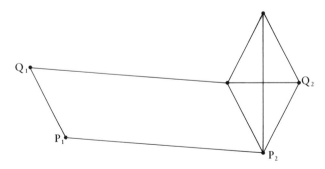

FIGURE 22.6

able fundamental concepts within an axiomatic theory? This is where Dingler's original idea comes in. These fundamental concepts are not simply any old fundamental concepts. Rather, they are much better characterized, using certain homogeneities, as spatial forms. What we mean here by homogeneity (Dingler spoke instead of symmetry) can be explicated using the fundamental concept of a plane. We start with the prescientific concept of a physical object or body. Surfaces form the boundaries of an object. These surfaces can be either flat or warped. What does that mean? Now, flatness can be characterized as our being unable to detect any variation. On a flat surface any place on the surface is exactly the same as any other place on the surface. That is also true, however, of the outer surface of a sphere. But in the case of a spherical surface the two sides are easily distinguished, namely, as either concave or convex.

This leads to the following Dinglerian characterization of planes using a principle of homogeneity: a flat surface of which all points (and the two sides) are indistinguishable.

If we are considering the two sides of a flat surface, then this reduces to the betweenness relation. We say that two points P_1 and P_2 lying outside of a plane E lie on the same side or different sides, if P_1EP_2 is valid or is not valid, respectively.

If we have planes at our disposal, then lines can be defined as the intersection of two planes; that is, the points that lie upon a line are those points that lie in both of two different planes.

Besides topological relations, which are not of interest here, there remain only parallelism and orthogonality. These can also be characterized using homogeneity. It may already be intuitively clear which homogeneities are required here. The parallelism of a plane E to a plane E' is characterized by all points of E being indistinguishable also in relation to E'. E' should lie symmetrically to all points of E.

For the orthogonality of a line l to a plane E with a base at P, the homogeneity principle says that all the lines of E that pass through P must be indistinguishable with respect to l.

I will set aside until later the logical analysis of these principles of homogeneity. For the moment I hope everyone will understand intuitively what these principles are about.

Because we have specified our fundamental concepts by requiring specific homogeneities, these concepts contain an element of the ideal. We might therefore confidently describe these concepts as geometrical ideas. If instead of "idea" we use the Aristotelian term "form," because it is better suited for things spatial, we can also call our fundamental concepts *homogeneous fundamental forms*.

The Foundational Problem of Geometry | 265

Of course, I do not want merely to draw conclusions from these new words. The designation "idea" or "form" is, however, also appropriate because in the case of ideas (not conceived here in strict accordance with Plato) we usually inquire into their realization. If we employ the Platonic and Aristotelian terms in a combination, we can then inquire more precisely into the material realization of our ideal forms: planes, parallelism, and orthogonality.

Our inquiry into their realization should serve only to clarify these homogeneous fundamental forms. The logical analysis and the deductions from the homogeneity principles to come later are independent of the issue of realization.

Historically, we can follow in a remarkable way the realization of the homogeneous fundamental forms by following the course of the Mesopotamian invention of bricks. At the beginning of the third millennium B.C. we still find Sumerian buildings made of dried, roughly shaped clay slugs (fig. 22.7). In the course of the first half of the millennium we then find, however, the following forms one after another.

FIGURE 22.7

1. The bottom is flattened (fig. 22.8). This is the realization of the fundamental form of the plane.

FIGURE 22.8

2. The sides are also flattened and are in fact perpendicular to the bottom and to each other (fig. 22.9). In this we already have the realization of all three fundamental forms.

FIGURE 22.9

3. The top is then also flattened parallel to the bottom (fig. 22.10). This

FIGURE 22.10

is the finished brick, the form of which (up to but not including the quantitative relations) is completely specified by the three fundamental forms.

As regards the homogeneous fundamental forms, a better material for intuition can hardly be imagined. Further consider that the realization of these forms is not of concern simply to primitive technology and of interest only to the historian. Even today the realization of the homogeneous fundamental forms in precision machinery is of decisive significance for our contemporary technology, most obviously in the production of optical apparatuses.

The art of lens grinding can in fact be traced back to the third millennium B.C. Schliemann found convex lenses ground from crystal in Troy I. Aquamarine lenses, the forerunners of our spectacles, are known from late antiquity on. Spherical surfaces were ground from glass as shown in figure 22.11. The longer one moves the upper piece of glass back and

FIGURE 22.11

forth, the more accurate the curvature becomes. Simultaneously, an ever more precise spherical surface is produced. This kind of grinding is still widely done by hand today, because the grinding motion must be as irregular as possible. In the United States we even find telescope making pursued as a hobby. To be sure, modern industry is able to replace the hand with expensive mechanical grinders; the grinding process remains, however, exactly the same. The process rests upon the homogeneity principle that characterizes spherical surfaces. Even the lenses of the telescopes with which the Einsteinian gravitation geometry was confirmed were produced in this way.

If you want to grind a plane rather than a spherical surface, then to make the surfaces indistinguishable you need three blocks to grind each other reciprocally. It is remarkable that this process has been employed only since the eighteenth century. You can easily figure out how parallel planes and right angles can be realized by grinding. In practice, of course, modern industry makes constant use of optical control processes; that is, it employs geometrical optics and thereby naturally employs Euclidean geometry as a theoretical tool.

From the glass grinder's perspective the most important questions concerning realization are the material questions, whereas for the geometer it is only a matter of the forms; in fact, every realization in a material will always fall short. The geometer looks—in Plato's sense—not toward the material realization but toward the idea itself, toward the pure form. It is only through knowledge of homogeneity that the intuition of realized forms becomes a pure intuition. Only the eye of the mind is capable of this intuition.

It is even more remarkable that the Kantian terms can also be easily brought together here with the Platonic, because Kant's philosophy is truly different *toto coelo* from Plato's. Nonetheless, both philosophers must have had intuitively before their eyes the same phenomenon of nonempirical geometry. Thus, in both cases geometry supplies at least a first entry into Kant and Plato. It supposedly was written over the entrance to Plato's Academy that one had first to know geometry before one would be allowed to enter. Most readers of Kant's critique of reason no longer satisfy this prerequisite, because the development of modern mathematics since the nineteenth century has led to an almost complete suppression of intuition. Therefore, it seems to me necessary explicitly to demonstrate these homogeneous fundamental forms.

I will now turn to the logical analysis of these forms—in particular, the possibility of deducing Euclidean geometry from the principles of homogeneity. To that end I want first to draw attention to the fact that the homogeneities of the fundamental forms are qualitative and not quantitative. In these homogeneities there is nothing to be measured; it is only with congruence that measurement becomes possible. An unevenness of surface is a quality. Plane is defined as the absence of unevenness. The same holds for parallelism and orthogonality. Quality systematically precedes quantity here.

Modern logic has made it generally known that mathematics is not the science of the quantitative. But, of course, that does not mean that vague and overlapping concepts are now allowed in mathematics. Qualities—in modern parlance, relations—are represented by symbols in mathematics, and the point of mathematics is then to operate only with these symbols according to unambiguous and specific rules. As a consequence of these rules—which also include logic—it all comes down again to counting, to calculi.

The requirement that all rules be made precise cannot be removed from geometry, if it is to be an exact science. So it seems essential to me that the previous formulation of the homogeneity principles be tightened so that these principles can serve as the basis for a mathematical theory.

What does it mean, for example, for all points of a plane to be indistinguishable? According to Leibniz, the indistinguishability of two objects amounts to their substitutability in all propositions. We then interpret the homogeneity of a plane relative to its points as follows: We require for every plane E and every one of its points P that from the truth of a proposition concerning E and P there follow the truth of those propositions that are produced when point P is replaced by any other given point P' of E. This homogeneity principle represents nothing more than a substitution rule. From every proposition $A(E,P)$ we are permitted to infer $A(E,P')$. If we write down the conditions of this substitution—namely, $P \ \varepsilon \ E$ and $P' \ \varepsilon \ E$—this rule reads

$$P \ \varepsilon \ E, P' \ \varepsilon \ E,, A(E,P) \Rightarrow A(E,P')$$

$A(E,P)$ here stands for an arbitrarily chosen proposition. For an axiomatic theory, that means that $A(E,P)$ is arbitrarily constructed from the theory's elementary propositions using logical operators. These elementary propositions are

$$P \ \varepsilon \ E, \ P \ \varepsilon \ l, \ P_1 E P_2, \ E_1 \| E_2, \ l \dashv E$$

In a particular instance, $A(E,P)$ can contain quantifiers and corresponding bound variables. We must require, however, that $A(E,P)$ contain no free variables except for E and P.

A substitution rule like the one above belongs to the metalanguage of a theory. But instead of a rule we can, when logic is brought in, also produce a so-called axiom schema:

$$P \ \varepsilon \ E \wedge P' \wedge A(E,P) \to A(E,P')$$

The $\wedge$ here stands for conjunction, and the $\to$ stands for the conditional, that is, for the if/then connective.

Without using formulas, instead of this axiom schema we could also say that all that holds of a point of a plane is also to hold of every other point. A plane is supposed to stand in the same relation to all of its points. This sameness is the discursive sense of "homogeneity."

Before I move on to the homogeneity principles for parallelism and orthogonality, I want to append yet another historical observation. Throughout the history of geometry there have always been attempts to define elementary concepts, in particular, lines and planes. Modern axiomatics considers such attempts unscientific. Nonetheless, these homogeneity principles supply a possible way to make sense of these attempts. It would be very remarkable if the notion of homogeneity had not been found earlier. In fact, even Euclid himself in the ὅροι of the first book said that a straight line is a line that lies equally to its points: ἐξ ἴσον χεῖται. Since antiquity this ἐξ ἴσον χεῖται has been a great problem for the commentators. It seems to me that with no great difficulty we can read from these words that Euclid wanted to give expression to the fact that a straight line shows no favor to any of its points, and that is indeed exactly what homogeneity makes more precise (what holds of one point also holds for all other points). In Leibniz, for example, homogeneity appears in the form: A plane is defined as a surface that divides space into two indistinguishable parts.

For orthogonality Euclid has under the ὅροι the sentence: The adjacent angles of an orthogonal are equal to each other. That is very closely related to the homogeneity principle formulated above. We can now write down this principle in a precise form using the following axiom schema:

$$P \varepsilon l \dashv E \wedge P \varepsilon h \subset E \wedge P \varepsilon h' \subset E \wedge A(E,l,h) \rightarrow A(E,l,h')$$

For orthogonals to be possible, the plane must also be in the same relation to its constituent lines. This latter homogeneity is what we find in Euclid as the definition of a plane.

I have been unable to find anywhere before Dingler a characterization of parallelism using a qualitative homogeneity principle. The Euclidean definition of parallels as straight lines never intersecting each other and the problematic of the parallel axiom have too strictly dominated the course of thought. Parallel means discursively "running next to each other." That nicely fits the requirement that a parallel plane—just as the plane itself—is to have the same relation to all points of the plane. So we have the following axiom schema:

$$P \varepsilon h \subset E_1 \wedge P'\varepsilon h' \subset E_1 \| E_2 \wedge A(E_1,E_2,P,h) \rightarrow A(E_1,E_2,P'h')$$

Axiom schemata like these which formalize the homogeneity principles— the indistinguishability of the sides of a plane will come later—are placed at the beginning of a theory. They must be supplemented by addition of existence axioms for the elementary concepts: For each plane E and each point P there exists a unique plane parallel to E through P, and similarly there exists a unique line through P and orthogonal to E.

If you prefer, it is not necessary to produce individual axioms here. We can instead incorporate these requirements into the theory's specifications for well-formed formulas. Moreover, I suspect that the uniqueness requirement will prove to be superfluous.

As regards the incidence and betweenness relations, that is, the topological relations, we must produce some further axioms. As examples I offer

$$P_1 E P_2 \wedge P \,\varepsilon\, E \rightarrow P E P_1 \vee P E P_2$$
$$P_1 E P_2 \wedge P_1 \,\varepsilon\, l \wedge P_2 \,\varepsilon\, l \rightarrow V_p.\, P \,\varepsilon\, l \wedge P \,\varepsilon\, E$$

Such topological axioms have no decisive implications for the problem of the a priori status of geometry, because no one would seriously claim that they are a posteriori judgments, that is, judgments that might eventually be disconfirmed by experience, particularly by physical measurements. Hugo Dingler's posthumous work *Aufbau der exakten Wissenschaften* (Munich: Eidos-Verlag, 1964) deals very thoroughly with these topological issues.

The Archimedean axiom plays a special role. It presupposes the concept of halving a line. This concept can be defined using the elementary forms in the following manner. Construct a rectangle above the line, draw diagonals, and then drop an orthogonal from the intersection of the diagonals to the line (fig. 22.12). Repeat this construction often enough,

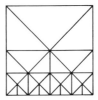

FIGURE 22.12

and you will arrive eventually at any arbitrarily chosen partial interval. How do we know that? In any event, this assertion is not refutable by experience. Neither is it logically necessary, as Vernoese has demonstrated. Archimedes said in one place that earlier geometers had obtained many indubitable theorems with this "supposition" (he was thinking of Eudoxus), and therefore he would use it too. Dingler justifies use of the Archimedean axiom on the basis of its being an expedient stipulation. If you wanted to insist on recognizing as different two points that could not be separated by halving, this distinction has nonetheless no practical use.

So we let it go. Logically, the stipulation concerning the indistinguishability of points is part of the definition of the concept of a point.

We are now faced with the exercise of formally deducing a complete axiom system for geometry from the homogeneity principles (that characterize the plane, parallelism, and orthogonality), certain topological axioms, and the Archimedean axiom. This exercise has not yet been completed. Fortunately, however, there is a school of geometers, so that a solution has already been sketched out clearly. I only want to use an example here to show how we can move from the homogeneity principles to the linkup with the usual geometrical axiomatics. To do this it is not necessary to carry everything out in strict formality.

For my example I choose the following theorem:
Every straight line that is orthogonal to a plane is also orthogonal to every parallel plane.
Proof. By presupposition, the figure consisting of parallel planes E_1 and E_2 and an orthogonal l_1 to E_1 (the point of intersection is P_1) has the same relation to all straight lines h of E_1 that pass through P_1. From P_1 let a perpendiular l_2 drop to E_2. Now, if l_2 is different from l_1, then together these straight lines define a plane, and this plane cuts E_1 in a specific line h through P_1. This contradicts homogeneity. Thus, l_2 and l_1 are the same; that is, l_1 is also orthogonal to E_2.

This theorem demonstrates clearly that only a Euclidean geometry, not a non-Euclidean geometry, results from the foundation I have presented here. This can also be explicitly demonstrated using the parallel axiom. Because "parallelism" occurs here as an elementary concept, we formulate the axiom as follows: Every straight line that intersects a plane also intersects every parallel plane.

If that can be demonstrated, then the plane parallel to E through a point $P \varepsilon E$ is the single noninteresecting plane through P. In that case, "nonintersecting" and "parallel" are equivalent concepts, as in the case of Euclid's definition.

I cannot demonstrate here the proof of this parallel axiom. I want only to note that it cannot be deduced from the homogeneity principle for parallelism alone. It could very well be that there are lines through every point of a plane that do not intersect a given parallel plane. Homogeneity would not thereby be contradicted.

The proof of the parallel axiom uses the theorem above and, in addition, the Archimedean axiom. We can show that the latter is in fact indispensable here. In such questions of independence versus dependence there are still many tasks for mathematical analysis before we will have finally traveled along the simplest road from the homogeneity principles of the elementary forms (plane, parallelism, and orthogonality) to a completely axiomatic theory for geometry.

As a concluding observation, I want to allow myself to look back over this theory as if it were already completed. Like every other axiomatic theory, mathematicians can handle it in a purely formal manner. But the theory is more than merely an axiomatic theory when we interpret its elementary concepts as elementary spatial forms, as we have done here following Dingler's lead in providing a foundation for the axioms and axiom schemata. Nonetheless, this interpretation of the elementary concepts does not turn the theory into a physical theory in the sense of modern physics. Modern physics understands itself to be an empirical science. Its propositions are hypotheses that, if not capable of confirmation by experience, are nonetheless capable of refutation. This clearly does not apply to the theory of homogeneous geometrical elementary forms. As regards the behavior of certain material bodies that are called "rigid" by an empirical physics, this geometry has absolutely nothing to say, for example, about whether such bodies would permit the production of very large parallel plates. Rather, the reverse is the case; this geometry defines when a body may be viewed as rigid. In Kant's sense, we can say that the geometry of homogeneous elementary forms supplies the conditions that make physical measurement possible in the first place. This geometry belongs to neither pure mathematics nor empirical physics. Rather, it represents the first step that leads beyond pure mathematics to physics. Therefore, I want to call this geometry a *protophysical* theory. Protophysics is not—as is the case in mathematics—a matter of constructing symbolic systems and operating formally with such self-generated symbols. It is much more the case that protophysics already intrudes into the reality immediately surrounding us. At any rate, it does not describe the behavior of real bodies; rather, it prescribes certain forms for these bodies. This discussion concerns geometry, and I shall not take up the question of whether parts of classical mechanics should also be counted as part of such a protophysics and, if so, which parts.

As regards the discussion of the peculiar position that geometry occupies between pure mathematics and empirical physics, I nonetheless want to direct your attention to the following. All that has been said to this point concerning geometry as the precondition of physics does not warrant the assertion that Euclidean geometry is a necessary precondition of *the* physics. The geometry of homogeneous elementary forms is indeed Euclidean geometry, and it makes possible a definition of rigid bodies and thereby makes possible a physics that measures things. But it does not follow from this that our modern physics is identical to the physics that this geometry makes possible. If we speak simply of physics, that is just a collective name for an immense number of past and present scientific endeavors. The word "physics" describes a region of human activity that is initially given to us only historically. If you want to assert that the propositions of a protophysics must necessarily be presupposed by "the" physics, then you need to have a systematic concept of physics. Such a concept can be precisely so defined that it presupposes the homogeneity principles of geometry—this idea was aggressively pursued by Dingler—but at the same time no one can be forbidden to believe that another physics, a physics quite close to the historically received physics, is at least possible.

It seems to me to be worthwhile to recommend that the discussion of these fundamental questions in the philosophy of science not be burdened by always taking several steps at once. In the present situation in which the discussion has bogged down over the possibility and necessity of the synthetic a priori, I would prefer to restrict myself to geometry. In geometry Dingler's investigations supply the beginnings of a geometry of homogeneous elementary forms that—although it is the old familiar Euclidean geometry—nonetheless can, I hope, reanimate the discussion simply because it does nothing more than indicate a possibility that until now, and indeed since Euclid, has been forgotten.

TWENTY-THREE

Geometry as the Measure–Theoretic A Priori of Physics

Modern physicists know that geometry can be traced back to Greek science. Thales (ca. 600 B.C.) is reckoned to be the first geometer. Euclid (ca. 300 B.C.) wrote the classic textbook on geometry.

With Descartes (1650) and the modern era came "analytic" geometry, that is, the arithmetization of all geometrical problems using a system of coordinates. Analytic geometry was then improved upon in the nineteenth century, first by Gauss and Riemann and then by Felix Klein (in the Erlanger Program, 1872).

Only thereafter was Euclid's "synthetic" method first brought up to the level of modern mathematics by Pasch (1882) and later by Hilbert (1899). But this synthetic method no longer played any part in physics, because the analytic method sufficed for all physical theories.

It was only with relativity theory and its "revolution" in concepts of space and time (i.e., geometry and kinematics) that philosophy of science became interested in clarifying the status of geometry as a theory that *precedes* physical theories. This *precedence* is articulated in expressions like "a priori theories" and "protophysical theories." An investigation of synthetic geometry is required to clarify the problem of geometry as a theory that prepares the ground for physics; analytic geometry is meaningful only on the basis of a demonstrated isomorphism between the regions of geometry and arithmetic.

Pasch and Hilbert looked on Euclidean geometry with the eyes of modern structural mathematicians—that is, as if Euclid's interest were to *describe* the *structure* of the region of geometrical objects (of points, lines, and planes in space) using a few simple propositions such that all the remaining propositions describing this structure could be proven using the "few simple propositions" (called *axioms*) alone. Pasch and Hilbert took incidence, ordering, and congruence as the elementary relations between geometrical objects. Hilbert chose—for plane geometry—fourteen axioms (I am ignoring the completeness axiom here) and then carried out the construction of geometry up to the proof of isomorphism with analytic geometry in such a way that at the same time the indispensability of each axiom was demonstrated. If any one of the fourteen axioms is dropped, the result is a pseudogeometry (a non-Euclidean geometry, as it is also called).

Now, you only need to read Book 1 of Euclid (at most, also the definitions in Book II) to see that in his textbook Euclid had something else in mind than did the modern structural mathematicians from Hilbert to the Bourbaki Group. Euclid constructed the forms of figures with a ruler and compass; that is, he constructed the forms out of straight lines (segments, rays, and lines).

The first three propositions of Book 1 reveal that congruence is defined using the *elementary forms* of straight lines and circles: Two segments from a point M are congruent, if the end points lie on a circle centered on M. If the segments do not emanate from a single point, then a construction must be performed (fig. 23.1). AB is then congruent with CD, if for

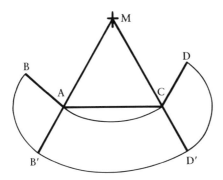

FIGURE 23.1

the equilateral triangle AMC B and B' lie on a circle with its center at A, D and D' lie on a circle with its center at D, and B' and D' lie on a circle with its center at M. Euclid's next proposition (1.4) is blemished in that this definition of congruence is not used. Consequently, Hilbert takes this proposition 1.4 (it is the proposition concerning the congruence of triangles with two sides and the enclosed angle being equal) as an axiom. After 1.4 Euclid makes no more demonstrative use of "bringing into coincidence" (ἐφαρμόζειν)—with the exception of 1.4 and 1.8, Euclid makes no mention of movement (translation of rigid bodies) in his geometry. The talk is rather of the forms of figures that can be constructed using lines and circles.

Since at least the time of Helmholtz it has been customary in physics to conceive "geometry" to be precisely a theory of the "free mobility" of rigid bodies (particularly, the mobility of measuring rods). This conception then led to Hilbert's axiomatization with congruence as an elementary relation (and thus led to the "transferring" of lines and the "inscribing" of angles).

So there remains the task of providing a foundation for geometry as a theory of the constructible forms of figures and doing so, as was first articulated by P. Janich,[1] on a level of rigor equal to that of Hilbert.

This need not mean that lines and circles must be taken as elementary forms (as is the case in Euclid). A simpler construction is obtained if we replace *circles* with *right angles*. In addition to the systematic reasons for this,[2] geometry's origin in Thales also speaks for such a substitution. Aside from *measuring rods* and *tapes,* which were used even in Mesopotamia and Egypt, the Greeks were the first to our knowledge to use *protractors* to measure angles. These protractors usually used the ancient division of the full circle into *six* parts with finer divisions using the sexagesimal system.

Since the French Revolution there has also been the newer division of the circle into *four* right angles R with finer divisions using the decimal system, for example, into 100 cR. A protractor measures angles by producing a circular sheet the edge of which has been marked off into equal parts using a measuring tape. Independently of the size of the sheet and the unit of measure used on the edge, the *relation* of the quantity measured for a given angle to the quantity measured for a right angle serves as the measure of the given angle. In the measuring of angles there is an *absolute angle* (R), whereas in the measuring of segments there is in contrast no *absolute length*. The "Thalesian elementary figure," as Oskar Becker has called it, consists of two lines intersecting at a point C, for which we then mark the bisectors with a protractor (figs. 23.2, 23.3). Because the angles to be bisected together equal $2R$, the bisectors are orthogonal to each other.

The four points on the edge of a protractor centered on C and the four lines that mark those points always result from the orthogonals constructed using the bisectors of any two intersecting lines.

1. P. Janich, in *Protophusik* (Frankfurt a. M.: Suhrkamp, 1976), ed. Gernot Boehme.
2. P. Lorenzen, *Theorie der Technischen und Politischen Vernunft* (Stuttgart: Reclam, 1978), p. 399.

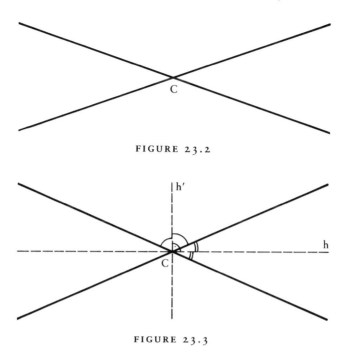

FIGURE 23.2

FIGURE 23.3

Thales' quadrilateral is defined to be a quadrilateral with the five orthogonalities indicated in figure 23.4. The circumscribed circle is called *Thales' circle*. According to Thales' *theorem*, every Thales' quadrilateral is a rectangle. The problem of geometry as a "theory" lies in "proving" such general conditional propositions. In particular, providing analytic geometry with a foundation requires proving the (Pythagorean) theorem concerning the sum of the squares: for $c = 2r$, $c^2 = a^2 + b^2$ holds. This theorem was also pursued by Euclid; he proved it in the last proposition of Book 1. The Thalesian elementary figure suggests that the "congruence" of segments be defined either as equal sides (in the Thales' quadrilateral) or as equal diagonals (in the Thales' quadrilateral). That would mean that two segments CA and CB that extend from a point C would be called diagonally equal, if a Thales' quadrilateral could be constructed from the free end points on the extended segments.

Two given segments are *congruent* if they are diagonally equal to two laterally equal segments.

This "congruence" also means "of equal size"—to distinguish it from protogeometrical congruence.

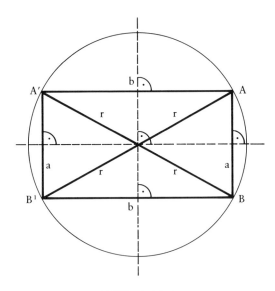

FIGURE 23.4

For two rays from a point C there can be constructed for every point A on the one ray a point B on the other ray, if (with a compass or even with only a ruler) we "transfer" the segment CA to CB (fig. 23.5). In this way the *angle bisector* can also be constructed (as the line that is orthogonal to AB through C).

If we were to proceed in the opposite direction from the constructability of the angle bisectors (which are defined as the mutually orthogonal lines that are orthogonals to the sides of a Thales' quadrilateral) through the midpoint, then we would thereby define the equality of diagonals.

These considerations lead to what I would like to call a Thalesian geometry that constructs the forms of figures using the fundamental

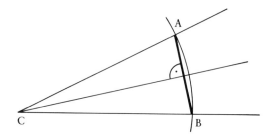

FIGURE 23.5

forms of the straight line and the right angle. As "axioms" we would posit the constructibility of connecting straight lines, orthogonals, *and* angle bisectors.

Two points suffice as the basis for all constructions in a plane. Protogeometrically, the two points are the two end marks on a measuring rod (say, the "original meter").

The technology of length measurement is a reproducible technology only insofar as it is scale-invariant: Pairs of points (serving as the basis for construction) cannot be distinguished by means of the figures constructed using them; these figures are identical in form ("similar").

The implicitly employed ordering of points on a line that occurs in Euclid is for Thalesian geometry also unproblematic. The ordering can be justified protogeometrically and is size-invariant. Therefore, the usual propositions concerning ordering—for example, the two Hilbert axioms

> Of three points on a straight line, there is exactly one between the other two.
> The Pasch axiom: A straight line that cuts a side but no corner of a triangle also cuts another side.

—can also be taken as "ordering axioms" for Thalesian geometry.

The situation then looks as follows. There are two domains of things to construct:

> points $A, B, C, \ldots$
> straight lines (lines, rays, lengths) $a, b, c, \ldots$

As fundamental relations we have:

> incidence $A|a$
> ordering ABC
> orthogonality $a \perp b$

(Instead of ABC we also write CBA; instead of $a \perp b$ we also write $b \perp a$.)

Using ordering in the usual way we define the segments AB and the two rays from a point A on a straight line a. We designate these rays using $A \rightarrow a$ and $A \leftarrow a$.

A segment AB is called a *fully oriented segment*, if one of the points, A for example, is marked *and* is on the line a that is orthogonal to AB through A of one of the two rays $A \rightarrow a$, $A \leftarrow a$.

As our basic axiom (1) it suffices to take the "existence" of two points. Axioms 2–3 are then the ordering axioms above.

To this beginning we then add *construction axioms*

4. For any two points A,B, there can be uniquely constructed a connecting straight line: $A \vee B$.
5. For every point A and every straight line a, there can be uniquely constructed an orthogonal through A to a: $A \dashv a$.
6. For any two rays $C \to a$, $C \to b$ from a point C, there can be uniquely constructed an angle bisector $(C \to a) \, \underline{\mathsf{Y}} \, (C \to b)$. (This is called the *principle of form*.)
7. Figures that are constructed in the same way out of two fully oriented segments are geometrically (i.e., on the basis of the fundamental relations) indistinguishable.

The following pages sketch out how to proceed with a proof of the sum-of-the-squares theorem using these seven "axioms" of a Thalesian geometry. Of Hilbert's fourteen axioms, two ordering axioms are retained. (Both of the other ordering axioms are provable using the construction axioms.) There remain ten axioms to be proved. Perusal of these axioms reveals that seven of them are trivial in a Thalesian geometry. The basic axiom and the constructibility of connecting straight lines provide the three plane axioms of incidence, and the first four congruence axioms result because, instead of congruence, "equivalence of length" is defined as the transitive closure of lateral equivalence and diagonal equivalence. There remains, then, only the congruence axiom for triangles (Euclid 1.4), the famous parallel axiom, and the Archimedean axiom.

We have to show that both of the first-named axioms—essentially by virtue of the principle of form—are provable in a Thalesian geometry. Then, of Hilbert's fourteen axioms there remains only the Archimedean axiom, which results trivially because according to the construction axioms, analytically speaking, as ratios of length only numbers of

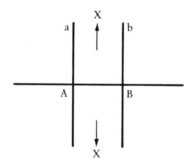

FIGURE 23.6

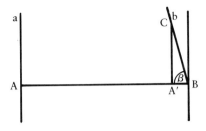

FIGURE 23.7

the Pythagorean series (which, given ξ, always contains $\sqrt{1+\xi^2}$) are constructible.

The parallel axiom is obtained as follows (see fig. 23.6).

PARALLEL THEOREM. Two lines a,b, have a common point, if and only if for a line c

$$a \perp c, \text{ but not } b \perp c$$

Proof. If for a line $a \perp c$ and $b \perp c$, then a and b do not cross. If the construction of the orthogonal rays $A \rightarrow a$ and $B \rightarrow b$ results in a common point (over the fully oriented base AB), then—according to the principle of form—this construction over the base AB with the orientation results in a common point for the rays $A \leftarrow a$ and $B \leftarrow b$, in contradiction of the uniqueness of the connecting straight line. Conversely, let there be over AB an orthogonal a and a line b with an angle β smaller than R being constructed (fig. 23.7). (Such an angle would be constructed from a right angle R by successive construction of angle bisectors.) On b, at least one point C may be constructed (e.g., with the orthogonal to b through A). The orthogonal $C \dashv (A \vee B)$ cuts AB at A'. Over $A'B$ can be constructed a right triangle with a hypotenuse angle of β. According to the principle of form, that is also valid for the segment AB.

The theorem of four right angles can be proven without using the theorems of congruence. If for two lines a third line is orthogonal to each, then any fourth line is also orthogonal to each if it is orthogonal to one of the first two. From this we get Thales' theorem. Thus the square (with orthogonal diagonals) can be constructed over any constructed segment, and finally—through arbitrary many iterations—a square lattice can be constructed (fig. 23.8). By halving, the square lattice can be made arbitrarily finer, and a measure of the segments that lie on the lattice line (relative to the lattice base) can be defined in the usual way. In other words, Cartesian coordinates for every (already constructed) point can be

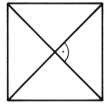 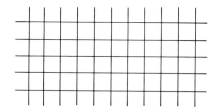

FIGURE 23.8

defined. If one chooses some other base for the lattice (e.g., one in which the previous base point 1,0 received the new coordinates $\lambda,0$, then the point with the old coordinates ξ,η receives the new coordinates $\lambda\xi,\lambda\eta$.

From this there immediately results the following (see fig. 23.9).

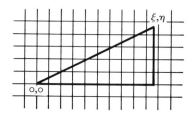

FIGURE 23.9

TANGENT THEOREM. For two right triangles that have the same hypotenuse angles, the ratio of the catheti will be the same.

Proof. We first construct a square lattice so that one of the catheti lies on the lattice line $\eta = 0$ and the hypotenuse angle has its vertex at point 0,0. The hypotenuse then has the end point ξ,η. Upon a second lattice base this point has the coordinates $\lambda\xi,\lambda\eta$. Thus there is a right triangle with the given hypotenuse angle that relative to the lattice base has its vertex at point $\lambda\xi,\lambda\eta$. According to the principle of form there is also such a triangle upon the first lattice base; that is, the point $\lambda\xi,\lambda\eta$ lies on its hypotenuse or its extension.

The ratio of the catheti is constant

$$\frac{\xi}{\eta} = \frac{\lambda\xi}{\lambda\eta}$$

From the tangent theorem the theorem of the sum of the squares can be proven in four steps (so the route through Hilbert's congruence axiom is a detour).

We next obtain the identity of the opposite angles of cut parallels because the opposite angles belong to the right triangles with catheti of identical lengths (fig. 23.10). In particular, it follows that the sum of the

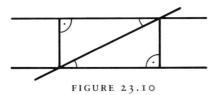

FIGURE 23.10

hypotenuse angles is R. We then obtain the height theorem. In a right triangle with height h and the segments of the hypotenuse cut by the heights p,q, $h^2 = pq$ (fig. 23.11).

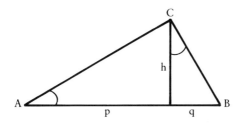

FIGURE 23.11

Proof. The angles at A and C are identical (because they both equal 90 when added to the hypotenuse angle B), thus

$$\frac{h}{p} = \frac{q}{h}$$

The theorem of the sum of the squares then follows for a right triangle ABC with catheti a,b and hypotenuse c (fig. 23.12). By construction of

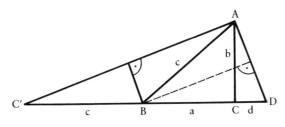

FIGURE 23.12

the angle bisector of B, we extend cathetus a to c beyond B to C'. The orthogonal to the second angle bisector through A cuts $B \vee C$ at D, and d is the segment CD. According to Thales' theorem $C'AD$ is a right triangle. Because $|BD| = C$, it follows that $d = c - a$ and, according to the height theorem, $b^2 = (c + a)(c - a)$ q.e.d.

With this theorem the coordinate transformation of a given square lattice (also "rotation") can be calculated. With this we have arrived at analytic geometry.

The *isomorphism theorem* is valid. For a Thalesian geometry the region of constructible elements is isomorphic with the region of Cartesian coordinate geometry over Pythagorean numbers.

Extension of the Pythagorean numbers (e.g., to the Euclidean numbers, which along with ξ always contain $\sqrt{|\xi|}$) can be accomplished as needed using constructive analysis. Like Hilbert's "completeness," that is an analytic, not a geometric, problem and will not be dealt with further here. For physics it is important only that with a Thalesian geometry we have available a *theory* of length measurement with which we can determine by reproducible means whether the standard of measurement used is "rigid," that is, whether when transported it always produces segments equal in length. A Thalesian geometry supplies the test criterion, because lines and angles must be checked for straightness and orthogonality, respectively.

Using the uniqueness theorems of protogeometry (for planes, lines, and right angles), we can justify these tests as reproducible processes. These test criteria for standard gauges are not in turn testable using length measurement (or other reproducible measurements). Therefore, we say that they are a theoretical measurement "a priori" for physics. "Geometry" is the name of the theory of the criteria for length measurement, that is, of the definition of identical size ("identical length") using the size-independent basic forms of the line and the right angle.

From a systematic perspective the objection to the Euclidean basic form of the circle is that after transport of a circle it must be possible to determine whether the size of the circle has changed. That is exactly what is supplied by the Thalesian definition of identical size for the radius.

So much then for geometry!

Were we all in agreement on the status of geometry as the theory of length measurement, then supposedly it would be easy to reach a consensus on kinematics—and then on collision mechanics and on dynamics from Newton to Einstein.

Without a consensus on geometry I can only sketch here—apparently dogmatically—how one can arrive at dynamics, if technical mastery of the movement of bodies is accepted as the goal of dynamics.

I use the term "kinematics" as it is defined in the 1970 Brockhaus encyclopedia, namely, as the name for that "part of mechanics in which only the movement of bodies is investigated *without* regard to the source of energy that causes the movements."

From a technological perspective kinematics is an important part of *kinetics*. A body has six degrees of freedom as regards mobility. Kinematic chains are built up from elementary pairs by means of which the mobility is limited to at least one degree of freedom. For example, a pair of planes has only three degrees of freedom, and a pair consisting of a nut and bolt has only one degree of freedom.

A chain with one degree of freedom is said to be "fixed."[3]

In order to determine the resulting velocities, we must define *clocks* as devices with which scale-invariant movement durations can be measured.

When we restrict ourselves to digital devices, an escapement with a counter, such a device is called a *clock* if it runs identically with all copies of the device started in arbitrarily chosen positions.

If G_1, G_2 are two clocks of one type and G_3, G_4 are two clocks of a different type, then, by definition for positions $\alpha, \beta, \gamma, \delta$, G_1^α, G_2^β run identically just as do G_3^γ, G_4^δ.

As regards the running relation between clocks, it follows from this that

$$\bigwedge_{\alpha,\beta,\gamma,\delta} G_1^\alpha : G_3^\gamma = G_2^\beta : G_4^\delta$$

so that, simply according to the logics of quantity and identity,

$$\bigvee_{\beta,\delta} \bigwedge_{\alpha,\gamma} G_1^\alpha : G_3^\gamma = G_2^\beta : G_4^\delta$$

and

$$\bigwedge_{\alpha',\gamma'} \bigwedge_{\alpha'',\gamma''} G_1^{\alpha'} : G_3^{\gamma'} = G_1^{\alpha''} : G_3^{\delta''}$$

The running relations for any two clocks are equal: We have a scale-invariant measure of duration (or "measure of time," as we also say). For the measurement of duration we assign to every lattice point a clock, all of which are synchronized with each other.

With length and duration we can, for example, measure velocity and acceleration.

Kinematics is the theory of *forms of movement* defined in this way. As the connection that leads from this kinematics (in which there is as yet no mass!) to dynamics, I propose *collision mechanics* for historical and systematic reasons.

3. Cf. W. Blaschke, *Ebene kinematik* (Munich, 1956).

Contrary to the actual history of science, in this proposal we abstract from the weight of bodies. This happened in practice, for example, in the development of billiards in the sixteenth century in Italy. Part of this theory are the laws of collisions that were formulated in the seventeenth century. Billiards represents an instance of completely elastic collisions, but John Wallis showed in 1670 how these could be reduced theoretically to inelastic collisions. In a completely inelastic collision between two particles along a line we have two velocities v_1, v_2 *before* the collision and a common velocity v *after* the collision.

This collision process can be reproduced technically in such a way that in the repetition of a collision of two particles (i.e., sufficiently small spheres) the velocity v becomes a function of v_1, v_2. The *collision law* is a function term S with $v = S(v_1, v_2)$. Relative to the earth it can be empirically determined that in the region of classical-technical velocities of bodies S describes a *homogeneous* function.

$$S(\lambda v_1, \lambda v_2) = \lambda S(v_1, v_2) \quad (\lambda > 0)$$

This result of measurement—we are now operating in empirical physics—permits us to define a "mass ratio" for two particles:

$$\frac{m_1}{m_2} \Leftrightarrow \lim_{v_1, v_2 \to 0} \frac{v_2 - v}{v - v_1}$$

We then measure the mass ratio for sufficiently small velocities. The limit definition is scale-invariant. From this definition there follows the classical law of momentum

$$m_1 v_1 + m_2 v_2 = (m_1 + m_2) v$$

for completely inelastic collisions. The quantity mv is here called "momentum."

The step to dynamics is completed by defining "force" as the way in which the momentum changes with time (t)—thus, *as if* the body were hit.

This leads to the Newtonian formulation in vector notation

$$M \frac{dV}{dt} = \sum_j K_j$$

in which we find suitable "force laws" that allow us to calculate the (observed or expected) change in momentum. Particularly in astronomy, with the law of gravitation

$$K \sim \frac{mM}{r^2}$$

the Newtonian formulation became self-evident for physics in the eighteenth and nineteenth centuries. Only toward the end of the nineteenth century did difficulties arise when *electrodynamics* could be brought into the Newtonian formulation using Maxwell's equations.

The theory of electricity began with the electrostatic definition of the charge ratio. By technical means one could produce charged bodies and measure the force of this "source" on charged test bodies—that is, the change in momentum. It can be demonstrated empirically that independently of the charge of the "source" and independently of the distance from the source" the ratio of the forces K_1, K_2 is everywhere dependent only on the two bodies used in the test.

This measurement process allows us to define a charge ratio

$$\frac{q_1}{q_2} \Leftrightarrow \frac{K_1}{K_2}$$

From the standpoint of the source, one could say that it produces at every point of space a "field strength" E, so that there acts upon a particle with a charge q a force

$$K = qE$$

In moving from particles to electrical fields one defines at every point a charge *density* ρ and calculates a force *density* k according to the Coulomb formula

$$k = \rho E$$

For moving charges we can calculate the force density for a given current density I (i.e., the charge density times the velocity) from the so-called magnetic flux density $*B$ using the Laplace formula $k = I \times *B$. Combining these, we obtain the Lorentz formula

$$k = \rho E + I \times *B$$

The vectors E and B (B is called the magnetic bivector and is the orthocomplement of $*B$) are calculated using Maxwell's equations and are dependent upon the charge density and the current density of the source.

Using the notation of the alternating differential forms, the two homogeneous equations are written as

$$dB = 0$$
$$dE + \dot{B} = 0$$

(the dot over B is Newton's notation for differentiation with respect to t), and for $\delta = *d*$ we then have two proportionalities

$$\delta B \sim I \quad \text{for } \dot{E} = 0$$

and

$$\delta E \sim \rho$$

With this we have the charge constancy

$$\delta I + \dot{\rho} = 0$$

As proportionality coefficients (for events occurring in a vacuum) we define

$$\delta B = \mu_0 I \quad \text{for } \dot{E} = 0$$

and

$$\delta E = \frac{1}{\varepsilon_0} \rho$$

On account of the charge constancy, for $\dot{E} \neq 0$ we must, as did Maxwell, put

$$\delta B = \mu_0 (I + \varepsilon_0 \dot{E})$$

The product $\mu_0 \varepsilon_0$ that appears here provides—surprisingly—the velocity of light c by means of the definition

$$\frac{1}{c^2} \Leftrightarrow \mu_0 \varepsilon_0$$

Using this definition, the nonhomogeneous Maxwell equations become

$$\delta B - \frac{1}{c^2} \dot{E} = \mu_0 I$$

$$\frac{1}{c^2} \delta E = \mu_0 \rho$$

The appearance of a velocity in this equation is an irritation for Newtonian mechanics. If, in order to measure space and time, we make use of a cubic lattice that we imagine to be firmly attached to our (earthly) laboratory, then we have—in this context we may ignore the rotation and revolution of the earth—an inertial system (i.e., a system in which the classical law of momentum is valid). A system of reference that moves uniformly relative to the system of the laboratory is then also an inertial system.

Lorentz and Poincaré recognized that the Maxwell equations can be formulated in a coordinate-invariant fashion, if we introduce a new "metric" (in Riemann's sense) for the uniform manifold of space–time points $x^1, x^2, x^3, x^4 = t$ that belong to the system of the laboratory. The *fundamental tensor* G_{mn} is defined so that the following is valid for the laboratory system

$$G_{mn} \begin{pmatrix} 1 & & & \\ & 1 & & o \\ & & 1 & \\ & o & & -c^2 \end{pmatrix}$$

Instead of the previous space vectors, we introduce here a *four-vector* by means of

$$\underline{B} = B + Edt$$

and

$$\underline{I} = I - c^2 \rho^{dt}$$

Instead of the Maxwell equations above, we obtain the *four-vector* equation

$$d\underline{B} = o$$
$$\delta \underline{B} = \mu_0 \underline{I}$$

Even this can be written more briefly still if we define a *four-vector potential* A by means of $d\underline{A} = \underline{B}$.

Thus there remains only one equation

$$\delta d\underline{A} = \mu_0 \underline{I}$$

All this can be viewed as a matter of mere mathematical elegance. Einstein saw, however, that these formulas imply a revision of the classical law of momentum. Instead of the Lorentz force density k, he used a *four-vector* force density $\underline{k}$, defined using $\underline{k} = \underline{I} \times {}^*\underline{B}$. In Minkowski notation, the Newtonian formulation

$$m \frac{dV}{d\tau} = \sum_j K_j$$

is revised so that there results a Newton-Minkowski formulation

$$m \frac{d\underline{W}}{d\tau} = \sum_j \underline{K}_j$$

for appropriate *four-sector* force laws $\underline{K}$.

The definition of a *four-vector* momentum $m\underline{W}$ and an "arc time," instead of the classical time coordinate t results directly from the Riemannian investigations concerning continuous manifolds with a metric defined by a fundamental tensor.

For the length σ of an arc,

$$d\sigma = \sqrt{\left(G_{mn}\frac{dx^m}{ds}\frac{dx^n}{ds}\right)}\, ds$$

is valid for every arc parameter s. The arc time τ is defined using $ic\tau = \sigma$, and the four-vector $\underline{W}$ (the four-vector momentum per mass) is defined by

$$\underline{W}^m = \frac{dx^m}{d\tau}$$

In the space coordinates ($m = 1, 2, 3$), $\underline{W}$ is distinguished from the classical velocity vector

$$V^m = \frac{dx^m}{dt}$$

by the famous factor

$$\beta_v = \frac{1}{\sqrt{\left(1 - \frac{v^2}{c^2}\right)}}$$

Because $\beta_0 = 1$, Newtonian dynamics is obtained as an approximation of Newton–Minkowski dynamics.

It seems to me to be confusing to claim that the kinematic velocity vector v is to be replaced by a "four-vector velocity" $\underline{W}$ as a consequence of a revision of *kinematics*.

It is rather—without going into classical kinematics—that a revision of *classical mechanics* leads to the replacement of classical momentum mV by a four-vector momentum. It is difficult to recognize this distinction, if one specifically calls collision mechanics "kinematics." Unfortunately, that is precisely what is usually done. We then lack a word for the a priori theory of the measurement of time and motion.

Einstein saw very early—that is, as early as 1907—that gravitation could *not* be handled using an appropriate four-vector force in the Newton–Minkowski formulation.

He had (as he himself said) the "best idea of his life" when he realized that no force at all was required in the treatment of weight in *homogeneous* gravitational fields. When the roofer falls from the roof, he has—as a physicist would see it—the good fortune of finding himself in an inertial system free from gravity.

All bodies fall with a constant acceleration. It is peculiar to "explain" this, then, by the fact that, on the one hand, "force = mass m times acceleration a" and, on the other hand, the force of gravity = mass m times the constant g. Because $m > 0$, it follows that $a = g$.

In *homogeneous* gravitation fields we do not need any four-vector force for the Newton–Minkowski formulation. *Before* application of the Newton–Minkowski formulation we change to a coordinate system that has an acceleration g relative to the system of the laboratory—the physicist lets himself fall from the roof (in thought!).

The *general theory of relativity* solves the problem for *inhomogeneous* gravitational fields by determining at every point of the field a coordinate transformation that produces locally a transformation to an inertial system.

Einstein and Hilbert found in 1915 the solution of first calculating a tensor field G_{mn} from the current density vector T_{mn} of the four-vector momentum using the field equation

$$R_{mn} - \frac{1}{2} G_{mn} \cdot R \sim T_{mn}$$

(R_{mn} and R are thereby determined by the contractions of the Riemannian tensor R^k_{mln}, determined in turn by G_{mn}.) Corresponding to the potential equation of electromechanics, G_{mn} is a "potential" of the requisite transformation to a local inertial system. This interpretation of general relativity theory was described by S. Weinberg in his book *Gravitation and Cosmology* (1972). According to this interpretation it is simply a misuse of geometrical language to speak of a revision of geometry.